高等学校中外合作办学适用教材

Introduction to Probability and Statistics

概率统计引论

陈建丽　申　敏　马树建　施庆生　编著

化学工业出版社
·北京·

内容提要

This book includes the probability of events, discrete random variables and their distribution, continuous random variables and their distribution, digital characteristics of random variables, law of large numbers and central limit theorem, sampling distribution, parameter estimation and hypothesis testing. All of the authors of this book have the background of visiting British and American university. The writing language is easy to understand, and the content has moderate difficulty. This book can be used for the teaching of probability and statistics courses of Sino-foreign cooperation projects and foreign student programs in universities of science and engineering (non-mathematics majors), as well as bilingual teaching of probability and statistics.

本书内容包括事件的概率、离散型随机变量及其分布、连续型随机变量及其分布、随机变量的数字特征、大数定律与中心极限定理、抽样分布、参数估计和假设检验。本书编著者均有英美访学背景，英文语言简单易懂，写作简约，内容难易适中，便于学习。本书可供理工科大学（非数学专业）中外合作办学项目和留学生项目的概率统计课程教学使用，也可供概率统计双语教学使用。

图书在版编目（CIP）数据

概率统计引论＝Introduction to Probability and Statistics：英文/陈建丽等编著．—北京：化学工业出版社，2020.7（2025.2重印）
高等学校中外合作办学适用教材
ISBN 978-7-122-36697-9

Ⅰ．①概… Ⅱ．①陈… Ⅲ．①概率论-高等学校-教材-英文②数理统计-高等学校-教材-英文 Ⅳ．①O21

中国版本图书馆 CIP 数据核字（2020）第 079584 号

责任编辑：郝英华　　装帧设计：刘丽华
责任校对：刘曦阳

出版发行：化学工业出版社（北京市东城区青年湖南街 13 号　邮政编码 100011）
印　　装：北京科印技术咨询服务有限公司数码印刷分部
710mm×1000mm　1/16　印张 13¾　字数 279 千字　2025 年 2 月北京第 1 版第 2 次印刷

购书咨询：010-64518888　　售后服务：010-64518899
网　　址：http://www.cip.com.cn
凡购买本书，如有缺损质量问题，本社销售中心负责调换。

定　　价：59.00 元　　**版权所有　违者必究**

Preface

Probability and statistics are a subject that studies the regularity of random phenomena. Because of the universality of random phenomena, the methods of probability theory and mathematical statistics almost infiltrate into the fields of natural science, technical science and economic management.

As an introduction to probability and statistics, in order to convey the importance and practicability of the course to the readers, this book first gives a brief introduction in Chapter 1 to its development history, research objects and applications in various fields of natural science and national economy. Then the part of probability theory (Chapter 2 to Chapter 6) and the part of statistics (Chapter 7 to Chapter 9) are discussed respectively. Probability theory, as the theoretical basis, mainly describes the probability of events, random variables and their distribution, numerical characteristics, the law of large numbers and the central limit theorem. The main contents of mathematical statistics include sampling distribution, parameter estimation and hypothesis test. The last part (Chapter 10) introduces how to use R software to solve the related calculation problems in probability and statistics.

In the process of writing this book, we strive to make the content easy to understand. We fully combine the characteristics of probability and statistics textbooks at home and abroad, weaken mathematical derivation in narration, and emphasize intuitive understanding. The selection of examples and exercises is combined with practice as far as possible, and we pay attention to the application of probability and statistics theories and methods in various fields. In addition, the discussion of this book tries to be in line with the thinking habits of the readers. The professional nouns in each chapter are annotated in the form of footnotes, which is convenient for readers to read and find.

In this volume, Chapter 1 is written by Shi Qingsheng, Chapter 2 and Chapter 6 are written by Ma Shujian, Chapter 3, Chapter 4 and Chapter 5 are written by Shen Min,

Chapter 7, Chapter 8, Chapter 9, Chapter 10 and Appendix are written by Chen Jianli. All the chapters are checked and revised by Chen Jianli. The overall framework of this book is determined under the guidance of Professor Shi Qingsheng, and he gives us a lot of valuable suggestions in the process of writing.

Due to the limit of our ability, there are inevitably shortcomings and mistakes in the book. We would appreciate any constructive criticisms and corrections from readers.

Chen Jianli, Shen Min, Ma Shujian, Shi Qingsheng

2020.6

前言

概率论与数理统计是研究随机现象规律性的一门学科。由于随机现象存在的普遍性，使得概率论与数理统计的方法几乎渗入到自然科学、技术科学以及经济管理等各领域中。

作为概率论与数理统计的入门教材，本书的编写首先通过第一章对它的发展历史、研究对象以及在自然科学、国民经济各领域中的应用进行简要介绍，以期向读者传递该课程的重要性及实用性。然后分别对概率论部分（第二章到第六章）和数理统计部分（第七章到第九章）展开论述。概率论部分作为理论基础部分，主要讲述事件的概率、随机变量及其分布、数字特征以及大数定律与中心极限定理。数理统计部分的主要内容包括抽样分布、参数估计和假设检验。最后部分（第十章）介绍了如何用 R 软件解决概率统计中的相关计算问题。

本书在编写过程中，努力做到通俗易懂，简详得当。充分结合国内外概率统计教材的特点，弱化数学推导，加强直观理解。例题和习题尽量联系实际，注重体现概率统计理论、方法在各个领域的应用。另外，本书的论述尽量做到符合读者的思维习惯，并对每一章中出现的专业名词以脚注的形式进行中文注释，便于读者阅读和查找。

参加本书编写的有施庆生（第一章）、马树建（第二、六章）、申敏（第三～五章）、陈建丽（第七～十章和附录），最后由陈建丽负责全书的统稿和定稿。施庆生教授指导确定了本书整体编写框架，并在编写过程中给出了许多宝贵意见。

本书的编写获得“江苏省第二批中外合作办学高水平示范性建设工程项目培育点：南京工业大学与英国谢菲尔德大学合作举办数学与应用数学（金融数学）专业本科教育项目”（苏教办外［2017］14 号）经费支持。

由于编者水平所限，书中难免存在错误和不足之处，敬请读者批评指正。

陈建丽、申敏、马树建、施庆生

2020.6

Contents

Chapter 1 Introduction

For readers who want to learn a new knowledge, they always want to have a general understanding of the subject. For this reason, before we talk about probability theory and mathematical statistics, we first give a brief introduction to its history, research object and its application in various fields of natural science and national economy.

1.1 The Origin of Probability Theory and Mathematical Statistics

The origin of probability theory is related to the problem of gambling. At that time, there was a heated discussion about the problem of "sharing gambling". Early in 1654, the gambler Merle asked the mathematician Pascal a question that had vexed him for a long time: "Two gamblers A and B agree to bet a number of games, who wins m games first, all bets belong to him. But when one of them wins $a(a<m)$ games and the other wins $b(b<m)$, the gambling stops. Assuming that A and B have an equal chance of winning each game, how should gambling be distributed reasonably?" In the process of studying this problem, some important basic concepts of probability theory have been conceived.

Three years later, in 1657, Huygens, a famous Dutch astronomical, physical and mathematician, tried to solve the problem by himself and wrote the result into a book, "The Calculation of the Game of Opportunity", which is the earliest book on probability theory.

With the development of science in the 18th and 19th centuries, it has been noticed that there is a certain similarity between certain biological, physical and social phenomena and opportunity games, and the probability theory has been applied to these fields. At the same time, it also greatly promotes the development of probability theory itself. The founder of making probability theory a branch of mathematics was the Swiss mathematician Bernoulli, who established the first limit theorem in probability theory, Bernoulli's law of large numbers. This is an extremely important result in the study of classical probability theory of equal possibility events, which illustrates that the frequency of Laplace systematically summarized the previous work, and wrote the book, "The Probability Theory of Analysis". He clearly gave the classical definition of probability, and introduced a more powerful analysis tool into the probability theory, which pushed the probability theory to a new stage of development. At the end of the 19th century, the Russian mathematicians Chebyshev, Markov, Liaponov and others established the general forms of the law of large numbers and the central limit theorem by using the analytical method, which scientifically explains why many random variables encountered in practice approximately have the normal distribution. Stimulated by physics at the beginning of the 20th century, people began to study stochastic processes. In this field, Kormogorov, Wiener, Markov, Sinchin, Levi and Feller have made outstanding contributions.

Mathematical statistics is the science of collecting data, analyzing data and making certain conclusions on the problems studied. All the data examined by mathematical statistics has random errors. This brings a kind of uncertainty to the conclusion based on this kind of data, and the quantification of the data has to rely on the concept and method of probability theory. The close relationship between mathematical statistics and probability theory is based on this point.

Statistics originates from the activity of collecting data. It is necessary to collect all kinds of relevant data from personal matters to the governance of a country. For example, in the ancient books of our country, there are many records about registered residence, land tax, military service, earthquake, flood, drought and so on. Of course, the activity of collecting and recording data itself is not equivalent to the science of statistics. The collected data need to be compared, sorted out, expressed in refined form, on the basis of which the things studied are quantitatively or qualitatively estimated, described and explained, and their possible development in the future is predicted. Only the theories and methods of doing these things can constitute the content of mathematical statistics, which is a kind of knowledge.

1.2 Random Phenomena and Random Trials

People often encounter two different kinds of phenomena in social practice and scientific experiments. One is the deterministic phenomenon, that is, under certain conditions, a certain result will inevitably occur or not occur. For example, a free fall always falls vertically to the ground at high altitude; Water boils at 100℃ at standard atmospheric pressure; A computer damaged by a system virus cannot complete a predetermined program, and so on. All these phenomena are deterministic phenomena. They express the inevitable relationship between conditions and results. Deterministic phenomena is very widespread. Advanced mathematics, linear algebra and so on are the mathematical disciplines that deal with this kind of phenomena.

The other is random phenomenon, that is, whether the result will occur under given conditions is unpredictable. For example, it is impossible to predict in advance whether a coin will be head up or back up when it falls on the table. It is impossible to predict what is the opening price of the Shanghai Composite Index on a stock trading day. It is impossible to know in advance the number of boys and girls born in a city on a certain day, and so on. All these phenomena are random phenomena. There is uncertain relationship between conditions and results.

In order to explore and study the regularity of random phenomena, it is necessary to carry out a series of experiments, including a variety of scientific experiments and observations of random phenomena. For example, flip a coin and observe whether the head up or the back up; measure the length of several parts of the same type; toss a dice and observe the number of points facing up, etc. All of them have the following common characteristics:

(1) The experiment can be repeated under the same conditions;

(2) There is more than one possible result of each experiment, and all possible results are known in advance;

(3) Which result will occur is unknown until the completion of the experiment.

Experiments with the above characteristics are usually referred to as random experiments. All experiments mentioned later in this book refer to random experiments. Random phenomena are often studied by observing random experiments.

1.3 Statistical Regularity of Random Phenomena

Through long-term repeated observation and practice, it is found that the random phenomenon shows a kind of uncertainty in one or more experiments, but when a large number of repeated experiments or observations are carried out under the same conditions, the random phenomenon shows a certain regularity. For example, it is uncertain whether a uniform coin is tossed head-up or face-up, but if you flip multiple times, the ratio of front to back is always close to 1 : 1, and the more times you flip it, the closer it is to this ratio. This inherent regularity in a large number of repeated experiments or observations is often referred to as the statistical regularity of random phenomena.

Engels pointed out: "On the surface, it's where contingency works. But in fact, this contingency is always dominated by the internal hidden laws, and the problem is just to discover these laws." Probability and statistics are mathematical disciplines that study and discover the statistical regularity of random phenomena.

1.4 Some Important Applications of Probability and Statistics

When science and technology are in a relatively rough stage, people often ignore random phenomena. With the development of science and technology, and the increasing demand for accuracy, people are unable to ignore the random phenomenon. French mathematician Laplace said: "The most important problems in life, the vast majority of which are essentially a matter of probability." Because of the extensiveness of random phenomena in nature, the methods of probability and statistics are infiltrating into almost all fields of natural science, technical science and economic management day by day. Common problems are:

(1) In industrial and agricultural production and scientific experiments, there are a wide range of problems, such as the estimation, inspection and control of product quality, which are very important to enterprise managers. All these problems belong to the application scope of probability and statistics.

(2) There are many random phenomena in hydrology, such as the annual discharge of a river, the maximum flood peak, the actual annual maximum water storage of a reservoir, and so on. The study of these problems is of great significance to the construction of dams and hydropower stations.

(3) There are a large number of random phenomena in biology and medicine. Probability and statistical methods are widely used in the spread and diagnosis of diseases, modern genetics and genetic engineering, and two marginal disciplines, "biological statistics" and "medical statistics", are formed.

(4) With the development of modern agriculture and the needs of disaster prevention and mitigation, the deterministic meteorological forecast in the past can no

longer adapt to the current economic development, because the meteorological problems are also random.

(5) Two basic variables should be considered in structural design, namely, the load effect S (the internal force of the structure caused by the load acting on the structure) and the resistance R (the ability of the structure to bear load and deformation). In the past, S and R are described as deterministic variables, and the "fixed value design theory" is used for structural design, but the reliability of this design is poor. However, in reality, S and R can not be deterministic because of the randomness of load and wind pressure. With the development of construction industry, the theories and methods of probability and statistics are introduced into structural design, which makes the design of building structure more accurate, safe and reliable.

(6) Many service systems, such as telephone communication, logistics scheduling, ship loading and unloading, machine maintenance, patient waiting, inventory control and so on, can be described by a kind of probability model, and the knowledge involved is queuing theory.

(7) The Monte Carlo method, which is based on the idea of using needle injection experiment to estimate the value of π in probability, is a calculation method based on probability and statistics. With the help of computer, this method plays an important role in the research of nuclear physics, surface physics, electronics, biology, polymer chemistry and artificial intelligence.

In addition, probability and statistics are widely used in automatic control, informatics, navigation, aerospace, big data, finance, insurance and so on.

Chapter 2
Basic Probability

As we all known, the concept of sets is the foundation of the probabilities. In this chapter, we firstly start with a review of set theory, and discuss its role in probability theory. We introduce the concept of a **measure**[1], which is a particular type of function that operate sets, and define probability to be a special case of measure. We then discuss how to assign numerical values to probabilities. Finally, the concept of **conditional probability**[2] is defined, which we could use to change our probabilities as our information changes.

2.1 Set Theory

Event and combinations of events occupy a central place in probability theory. The mathematics of events is closely tied the theory of sets. Set theory is found by Georg Cantor in 1874 which is commonly used as a foundational system, although in some areas—such as algebraic geometry and algebraic topology—category theory is thought to be a preferred foundation. Here we mainly introduce the basic concepts and notation.

2.1.1 Sets, Elements, and Subsets

Definition 2.1 Set[3]

A **set** is a collection of objects, with each object referred to as an **element** or a **member** of the set.

These objects are called elements of the set and they can be of any kind with

[1] measure：测度
[2] conditional probability：条件概率
[3] set：集合

any specified properties. We may consider, for example, a set of numbers, a set of points, a set of functions, a set of animals, or a set of mixture of things. The capital letters A, B, C, Φ, Ω, $\cdots$ shall be used to denote sets and lower case letters a, b, c, φ, ω, $\cdots$ denote their elements. We can take some examples as following.

(1) S_1, a set of some colors.

$S_1 = \{$yellow, red, blue, green, white, black, orange, brown, grey, pink$\}$.

(2) N, the natural numbers.

$$N = \{0, 1, 2, 3, 4, \cdots\}.$$

We find that S_1 is a **finite set**[1], as it only has 10 members, and N is an **infinite set**[2], as there are infinitely many natural numbers.

We use the notation

$$\text{red} \in S_1,$$

to mean that red is a member or element of the set S_1. You should be able to guess what is meant by

$$\frac{1}{3} \notin N.$$

It means that $\frac{1}{3}$ is not the element of the set N.

We represent some sets using interval notation.

Definition 2.2 Open intervals[3], closed intervals[4], and half-open intervals[5]

Open, **closed**, and **half-open** intervals are defined as follows.

Open interval: $(a, b) = \{x \in R; a < x < b\}$,

Closed interval: $[a, b] = \{x \in R; a \leqslant x \leqslant b\}$,

Half-open interval: $(a, b] = \{x \in R; a < x \leqslant b\}$,

or $[a, b) = \{x \in R; a \leqslant x < b\}$.

Here you should read the semicolon to mean "such that". Note that we may write, for example, $(a, +\infty)$, but we do not write $(a, +\infty]$.

Definition 2.3 Subset[6]

For any two sets A and B, we say that set A is called a subset of B if every

---

[1] finite set：无限集合

[2] infinite set：有限集合

[3] open intervals：开区间

[4] closed intervals：闭区间

[5] half-open intervals：半开半闭区间

[6] subset：子集

element of A is also an element of B, and this is represented symbolically by

$$A \subseteq B \text{ or } B \supseteq A.$$

This does not rule out the possibility that A and B are the same set. If there is at least one element of B that is not in A, and we write

$$A \subset B.$$

In this case, we say that A is a **proper subset**[1] of B.

【Example 2.1】 If $R = \{2, 4, 6, 8, 10, 12, \cdots\}$ is the set of all even numbers, then $R \subset N$.

Definition 2.4 Empty set[2] and Space[3]

We define the **empty set** to be the set that does not contain any elements. We denote the empty set by ϕ. The "largest" set containing all elements of all the sets under consideration is called **space**, and it is denoted by the symbol S.

2.1.2 Set Operation: Union, Intersection, Complement and Set Differences, Exclusive and Opposite

We start with a **space** S, which lists all the elements we wish to consider for the situation at hand. We can then perform operations on subsets of S. We are primarily concerned with addition, subtraction, and multiplication of these sets.

Definition 2.5 Union or sum of two sets[4]

The **union** or **sum** of two sets A and B, written $A \cup B$ is the set of all elements that are either in set A, or in set B, or in both. We write

$$A \cup B = \{x \in S; x \in A \text{ or } x \in B \text{ or } x \in \text{ both } A \text{ and } B\}. \quad (2.1)$$

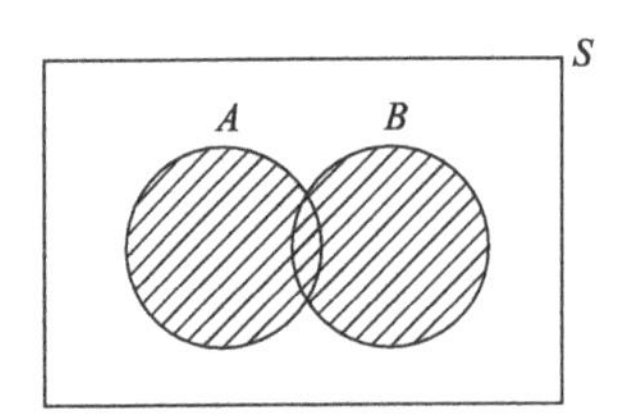

Figure 2.1 A Venn diagram, with the shaded region showing $A \cup B$.

We can visualize this using a *Venn diagram* (see Figure 2.1). The rectangle represents the universal set, the two circles represent the subsets A and B, and the shaded area represents the set $A \cup B$.

Definition 2.6 Intersection or product of two sets[5]

The **intersection** or **product** of two sets A and B (see Figure 2.2), written $A \cap B$ is the set of all elements that are both in the set A and in the set B. We write

---

[1] proper subset：真子集

[2] empty set：空集

[3] space：全集

[4] union or sum of two sets：集合的并或和

[5] intersection or product of two sets：集合的交或积

$$A \cap B = \{x \in S; x \in A \text{ and } x \in B\}. \quad (2.2)$$

If there are no elements in S that are both in A and in B, we say that A and B are **disjoin (exclusive)**, and we write

$$A \cap B = \phi.$$

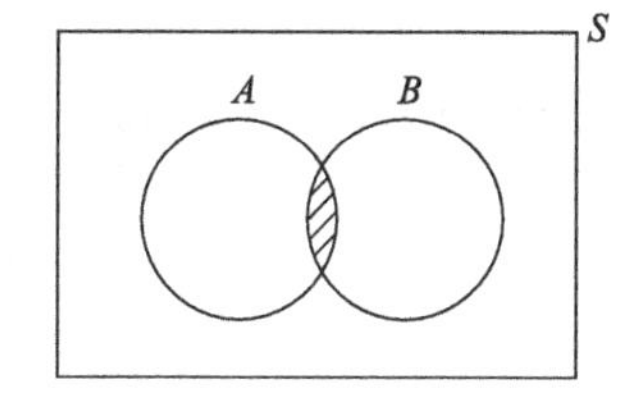

Figure 2.2 A Venn diagram, with the shaded region showing $A \cap B$.

Definition 2.7 Complement of a set[1]

The **complement** of a set A (see Figure 2.3), written as $\overline{A}$, is the set of all elements that are in S, but not in A. We write

$$\overline{A} = \{x \mid x \in S; x \notin A\}. \quad (2.3)$$

Alternative notation: $\overline{A}$ is also written as A^C and A'.

Note that

$$\phi = \overline{S}. \quad (2.4)$$

Definition 2.8 Set difference[2]

The **set difference** (see Figure 2.4), written as $A \setminus B$, is the set of all elements that are in A, but are not in B. We write

$$A \setminus B = \{x \in S; x \in A \text{ and } x \notin B\}. \quad (2.5)$$

Alternative notation: $A \setminus B$ is also written as $A - B$.

Note that

$$A \setminus B = A \cap \overline{B}. \quad (2.6)$$

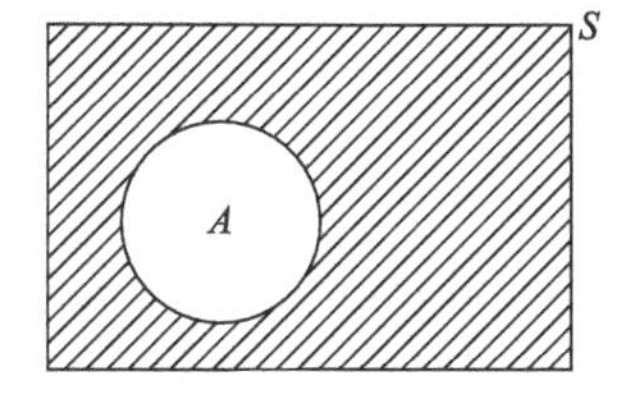

Figure 2.3 A Venn diagram, with the shaded region showing $\overline{A}$.

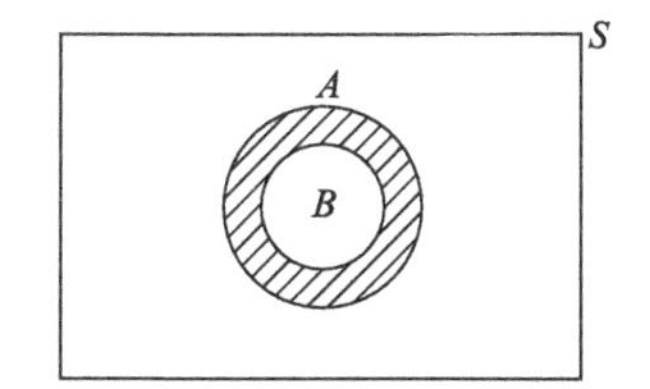

Figure 2.4 A Venn diagram, with the shaded region showing $A \setminus B$.

Definition 2.9 Symmetric difference[3]

The **symmetric difference**, written as $A \triangle B$, is the set of all elements that are either in A, or are in B, but are not in both A and B We write

---

[1] complement of a set：集合的补集

[2] set difference：集合的差

[3] symmetric difference：对称差

$$A\triangle B=\{x\in S; x\in A \text{ or } x\in B \text{ and } x\notin A\cap B\}. \tag{2.7}$$

【Example 2.2】 Suppose we have

$$S=[0,9]=\{x\in R; 0\leqslant x\leqslant 9\},$$
$$A=(1,5)=\{x\in R; 1<x<6\},$$
$$B=[3,7]=\{x\in R; 3\leqslant x\leqslant 7\}.$$

Then

$$A\cup B=\{x\in R; 1<x\leqslant 7\},$$
$$A\cap B=\{x\in R; 3\leqslant x<6\},$$
$$\overline{A}=\{x\in R; 0\leqslant x\leqslant 1; 6\leqslant x\leqslant 9\},$$
$$A\backslash B=\{x\in R; 1<x<3\},$$
$$A\triangle B=\{x\in R; 1<x<3; 6\leqslant x\leqslant 7\}.$$

2.1.3 Experiments, Sample Spaces, and Events

We now discuss the role of set theory in probability. We consider uncertainty in the context of an **experiment**, where we use the word experiment in a loose sense to mean observing something in the future, or discovering the true status of something that we are currently uncertain about. In an experiment, suppose we want to consider how likely some particular outcome is. We might start by considering what all the possible outcomes of the experiment are. We can use a set to list all the possible outcomes.

Definition 2.10 Sample space[1]

A **sample space** is a set which lists all the possible outcome of an 'experiment'.

【Example 2.3】

(1) We will observe how many matches a basketball team wins during a CBA season (46 matches). The sample space is

$$S_1=\{0,1,2,3,\cdots,46\}.$$

(2) A sample of blood, and we find out what blood group it belongs to. The sample space is

$$S_2=\{\mathrm{A},\mathrm{B},\mathrm{AB},\mathrm{O}\}.$$

(3) We observe the number of miles a new car is driven before its first breakdown. The sample space is

$$S_3=[0,+\infty)=\{x\in R; x\geqslant 0\}.$$

Note: Though we know the car can't travel infinitely far, writing the sample space as $[0,+\infty)$ rather than $[0,a]$ for some finite a, and make us not to worry about the value of a choose.

---

[1] sample space：样本空间

Definition 2.11 Event[1]

An **event** is a subset of a sample space. If the outcome of the experiment is a member of the event, we say that the event has **occurred.**

【Example 2.4】 If $R=\{0,1,2,3\}$, then $R\subset S_1$. The event R is the event that the team wins no more than 3 matches next season.

Given a sample space S and two subsets/events A and B, the set operations including union, inter-section, complement and difference all could define related events.

- The event $A\cup B$ is the event that either A or B occurs, or both occur.
- The event $A\cap B$ is the event that both A and B occur at the same time.
- The event $\overline{A}$ is the event that A does not occur.
- The event $A\setminus B$ is the event that A occurs, but B does not.

Let's clarify the difference between an "outcome" and an "event". An outcome is an *element* of the sample space S, whereas an event is a *subset* of S. Consider the following example.

Suppose Team A plays football with Team B in a match. The sample space is

$$S=\{\text{Team } A \text{ wins}, \text{Team } B \text{ wins}, \text{draw}\}.$$

- Before the match, you can bet on an *event*, for example, "Either Team B wins or draw".
- Before this match, a news reporter will tell you the *outcome*, for example, "Team A won". The news reporter would not report the event "either Team A won or the match was drawn last night".
- If the observed outcome belongs to the event "either Team A wins or draw", the event has occurred, and you have won your bet.
- An event could be a single outcome (you could bet on "Team B wins").

2.2 Set Functions[2]

2.2.1 Boolean Algebras[3]

Shortly, we will give a mathematical definition of "the probability of an event". However, it turns out to be difficult to do for all possible events (subsets) of all possible sample spaces, so we restrict our attention to sets that are members

---

[1] event：事件

[2] set functions：集合函数

[3] boolean algebras：布尔代数

of *Boolean algebras*, which we define shortly.

First, we consider the following example, suppose the two sets

$$S=\{a, b, c\},$$
$$S_1=\{\{a\},\{b, c\}\}.$$

S is a set with three elements: a, b and c. The set S_1 is a set with *two* elements. *Each element* of S_1 is itself a set: the set $\{a\}$ and the set $\{b, c\}$. In this example, the elements of S_1 are also subsets of the set S. Given any set S, a **Boolean algebra** is a set of subsets of S that satisfies the following two conditions.

Definition 2.12 **Boolean algebra**

Given a set S, a **Boolean algebra**, denoted by $\boldsymbol{B}(S)$ is a set of subsets of S that satisfies the following conditions.

(1) If $A \in \boldsymbol{B}(S)$, then $\bar{A} \in \boldsymbol{B}(S)$.

(2) If $A \in \boldsymbol{B}(S)$ and $B \in \boldsymbol{B}(S)$, then $A \cup B \in \boldsymbol{B}(S)$.

In the example above, we can see that S_1 is *not* a Boolean algebra. S_1 does not satisfy the condition (1) and (2).

Theorem 2.1 **Properties of Boolean algebras**

(1) If $A \in \boldsymbol{B}(S)$ and $B \in \boldsymbol{B}(S)$, then $A \cap B \in \boldsymbol{B}(S)$.

(2) $\phi \in \boldsymbol{B}(S)$ and $S \in \boldsymbol{B}(S)$

Examples of Boolean algebras

(1) For any set S, we first define the *power set* of S to be the set whose elements are all the subsets of S. The power set is a Boolean algebra. We denote the power set of S by $P(S)$.

For example, if

$$S=\{a, b\},$$

then

$$P(S)=\{\phi, \{a\}, \{b\}, \{a, b\}, S\}.$$

You can see any power set of S will include the empty set ϕ, and S itself.

(2) If our experiment produces a continuous quantity, our sample space may be the whole real line R, or at least some subset of it. We now construct a Boolean algebra based on R.

Following the definition of an open interval in Definition 2.12, we define $I(R)$ to be the smallest Boolean algebra containing all open intervals (a, b), where $-\infty \leqslant a \leqslant b \leqslant \infty$.

Theorem 2.2 **Properties of a broken line**

If $I(R)$ is a Boolean algebra and $(a, b) \in I(R)$, then

(1) $\overline{(a, b)} \in I(R)$.

(2) $[a, b] \in I(R)$.

(3) $\{a\} \in I(R)$.

From condition (2) in Definition 2.12, we can see that unions of single points and open/closed/half-open intervals are also members of $I(R)$.

2.2.2 Measures

Given our sample space, we now want to quantify how likely various outcomes or events are within the sample space. We need a function that we could apply to a set, and we make use of a particular type of function, known as a **measure.**

Following Applebaum (2008), we have made a simplification here. Instead of using a **Boolean algebra**, the formal definition of a measure uses what is called σ-algebra, where condition (2) in Definition 2.12 is extended to countable unions of sets, and the property in equation (2.8) is extended to $m\left(\bigcup_{i=1}^{\infty} E_i\right) = \sum_{i=1}^{\infty} m(E_i)$ for disjoint set $E_1, E_2, \cdots$. The simpler Definition 2.13 will do for this module.

Definition 2.13 **Measure**

A (finite) **measure** is a function that assigns a non-negative real number $m(A)$ to every set $A \in \boldsymbol{B}(S)$, that satisfies the following condition: for any two disjoint sets A and B in $B(S)$,

$$m(A \cup B) = m(A) + m(B) \tag{2.8}$$

Informally, a helpful example of a measure to consider is "total area" of sets drawn on a Venn diagram. Areas must be non-negative, and the total area of two disjoint sets is the sum of the areas of each set.

Note that we can extend the condition in equation (2.8) to finite unions of disjoint sets. If three sets A, B and C are all mutually disjoint, if we define $D = B \cup C$, then

$$\begin{aligned} m(A \cup B \cup C) &= m(A \cup D) \\ &= m(A) + m(D) \\ &= m(A) + m(B \cup C) \\ &= m(A) + m(B) + m(C) \end{aligned}$$

So, for finite unions of disjoint sets $E_1, E_2, \cdots, E_n$, $m\left(\bigcup_{i=1}^{n} E_i\right) = \sum_{i=1}^{n} m(E_i)$.

2.2.3 Examples of Measures

1. Counting measure

For a finite set S and Boolean algebra $\boldsymbol{B}(S)$, the counting measure of a set A gives the number of elements of that set. For example,

$$m(\{\text{red}, \text{yellow}\}) = 2.$$

We sometimes represent the counting measure by the symbol #:

$$\#(\{\text{red}, \text{yellow}\})=2.$$

2. Lebesgue measure

Let S be the real line R, and let $\boldsymbol{B}(S)=I(R)$. Then

$$m([a,b])=b-a.$$

The Lebesgue measure gives the length of an interval.

Theorem 2.3 **Properties of measures**

Let m be a measure on $\boldsymbol{B}(S)$ for set S, and A, $B\in\boldsymbol{B}(S)$.

(1) If $B\subseteq A$, then

$$m(A\backslash B)=m(A)-m(B). \tag{2.9}$$

(2) If $B\subseteq A$, then

$$m(B)\leqslant m(A). \tag{2.10}$$

(3) $m(\phi)=0$.

(4) $m(A\cup B)=m(A)+m(B)-m(A\cap B)$. (2.11)

2.2.4 Measures on Partitions of Sets[1]

The **partitions** of one set is given in the following.

Definition 2.14 **Partition**

A **partition** of set S is a collection of sets $\varepsilon=\{E_1, E_2, \cdots, E_n\}$ such that

(1) $E_i\cap E_j=\phi$, for all $1\leqslant i$, $j\leqslant n$ with $i\neq j$;

(2) $E_1\cup E_2\cup\cdots\cup E_n=S$.

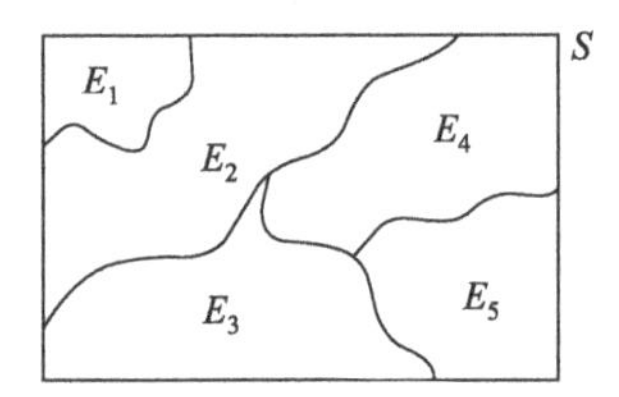

Figure 2.5 An example of a partition with 5 sets.

Here $\{E_1, E_2, \cdots, E_n\}$ are *mutually exclusive* and *exhaustive*. If S is a sample space, one of the events in ε *must* occur, but no two different events could occur at the same time.

Figure 2.5 shows an example of a partition with 5 sets.

From equation (2.8), we have

$$m(S)=\sum_{i=1}^{n}m(E_i). \tag{2.12}$$

2.3 Probability as Measure

Now we could assume S is one sample space, $B(S)$ is a Boolean algebra, and

---

[1] partitions of sets：集合的划分

P is a measure on $B(S)$. P is a *probability* (measure) if $P(S)=1$. $(S, B(S), P)$ is regarded as one **probability space**[1]. The three parts of a probability space are

S: the set of all possible outcomes in one 'experiment',

$\boldsymbol{B}(S)$: the set of all possible events,

P: a function that gives the possibility for every possible event.

2.3.1 Properties of Probability

According to the definition and the properties of measures, we could deduce some properties of probabilities. In the following, we will provide some properties of probability. For probability space $(S, B(S), P)$, with $A, B \in B(S)$,

(1) $P(S)=1$.

(2) $0 \leqslant P(A) \leqslant 1$.

(3) $P(\phi)=0$.

(4) $P(\bar{A})=1-P(A)$.

(5) $P(A \cup B)=P(A)+P(B)-P(A \cap B)$.

(6) If $A \cap B=\phi$, then $P(A \cup B)=P(A)+P(B)$.

(7) If $B \subseteq A$, then $P(A \backslash B)=P(A)-P(B)$ and $P(B) \leqslant P(A)$.

(8) If $\varepsilon=\{E_1, E_2, \cdots, E_n\}$ is partition of S, then $P(S)=\sum_{i=1}^{n} P(E_i)$.

For property (5), we could extend it to the finite set $A_1, A_2, \cdots, A_n$.

$$P(A_1 \cup A_2 \cup A_3 \cup \cdots \cup A_n)=\sum_{i=1}^{n} P(A_i)-\sum_{1 \leqslant i<j \leqslant n} P(A_i A_j)+\sum_{1 \leqslant i<j<k \leqslant n} P(A_i A_j A_k)-\cdots+(-1)^k P(A_1 A_2 \cdots A_n).$$

Each of these properties could be proved using the basic definition and properties. Note that we have assumed properties (1), (6), and that $P(A) \geqslant 0$ for all $A \in B(S)$ in the definition of a (probability) measure. Everything else follows as a consequence of these assumptions. The properties (1), (2) and (6) are sometimes presented as the three axioms of probability.

【Example 2.5】 Suppose team A plays team B in one test. We have

$$S=\{A \text{ win}, B \text{ win}, \text{Draw}\}.$$

If we consider each set A that belongs to $B(S)$, the probability measure P can be any function that satisfies the following together with the constraints $0 \leqslant p_1, p_2, p_3 \leqslant 1$,

---

[1] probability space：概率空间

A	$P(A)$
ϕ	0
A win	p_1
B win	p_2
Draw	p_3
A win or draw	p_1+p_3
B win or draw	p_2+p_3
A win or B win	p_1+p_2
A win, draw or lose	$p_1+p_2+p_3$

and $p_1+p_2+p_3=1$.

2.4 Assigning Probabilities

Using measure theory, we have established some basic rules for measuring uncertainty using probability. We have defined a scale: probability ranges from 0 (for an event that we are certain will not occur) to 1 (for an event that we are certain will occur). We have also chosen as an axiom an addition rule: if two events A and B cannot both occur, then the probability that either A or B occurs is the sum of the probability that A occurs and the probability that B occurs. But this is not enough to be able to use probability theory in the real world! We have *not* said anything about how we might choose the value of a probability in practice. In Example 2.5, the probability measure was determined by p_1 and p_2 (with $p_3=1-p_1-p_2$), but *any values* of p_1 and p_2 were allowed, as long as $0\leqslant p_1, p_2\leqslant 1$, and $p_1+p_2\leqslant 1$.

2.4.1 Classical Probability[1] Based on Symmetry

Definition 2.15 Classical probability model

The experiment is called classical probability model which satisfy:

(1) the total number of the event possible outcomes is infinite;

(2) the probability of each event outcomes is same

The events satisfy the condition (1) and (2) are called same probability events. Under this condition, we would use the **classical** approach to describe the probability of the event.

---

[1] classical probability: 古典概率

Definition 2.16 **Classical probability**

The probability of the event A is calculated as the number of outcomes in which the event occurs, divided by the total number of possible outcomes.

$$P(A)=\frac{\#(A)}{\#(S)}. \tag{2.13}$$

Here "#" means the number of the event outcomes.

In the example 2.5, we could assume the symmetry of the test and the events occur at the same probability.

$$P(A\ \text{win})=P(B\ \text{win})=P(\text{draw})=\frac{1}{3},$$

So that $p_1=p_2=p_3=\frac{1}{3}$. In effect, we have chosen P to be the counting measure, divided by the number of elements in the sample space.

2.4.2 Counting Methods for Classical Probability: Permutations and Combinations

To make use of classical probability, we have to count the number of outcomes in which the event occurs. Some useful results are as follows. Firstly, recap that for a positive integer n,

$$n!=n\times(n-1)\times(n-2)\times\cdots\times2\times1.$$

Here $n!$ is read as *n factorial*. The simple observation that $n!=n\times(n-1)!$ is very useful. To be consistent with this, when $n=0$, we define $0!=1$.

Multiplication principle[1]

Multiplication plays a fundamental role in counting problems. If an operation is composed of several different stages, and each stage can be played in a number of different ways, then the number of ways in which the overall operation could be described is the *product* of these numbers.

We shall see this principle in different types of counting problem. Here we give some examples.

【Example 2.6】 Counting methods

(1) When choosing a ticket from City A and City B. And City C is located between City A and City B. There are 5 kinds of tickets from City A to City C and 7 kinds of tickets from City B to City C. As we know, the total numbers of tickets from City A to City B is 35 ($35=5\times7$).

(2) A group of 256 work colleagues organize a weekly prize draw, each pay-

[1] multiplication principle：乘法原理

ing 1 per week. Each week three prize winners are drawn, using 256 balls, numbered 1 to 256, to be drawn from a bag and not replaced. The first wins £100, the second £50 and the third £30. How many different outcomes are there? ($256\times255\times254=16581120$)

Some general principles

Case A: with repetition, order matters

If we want to get r successive choices from a set of n elements, allowing repetition but with order mattering, the number of outcomes is n^r. We start to analyze this problem. Firstly, we choose the r elements from n one at one time. The first chosen one may be in n ways, the second may be chosen in n ways, and so on, the number of choices at each of the r stages always being n. By the multiplication principle, the total number of possible ways of making the overall choice is

$$n\times n\times n\times\cdots\times n\times n=n^r.$$

The number of outcomes $(5,5,5),(5,5,4),\cdots,(1,1,1)$ when a die is rolled three times is $5^3=125$.

Case B: no repetition, order matters (Permutations)

When we want to get r successive choices from a set of n elements allowing no repetition and with order mattering, the number of outcomes is

$$n(n-1)(n-2)\cdots(n-r+1)=\frac{n!}{(n-r)!}.$$

This number is named as P_n^r, the P referring to the fact that choices of this kind are called *permutations* of r elements from n. In particular, taking $r=n$, the number of ways of putting n objects in order is $n!$.

Theorem 2.4 Number of ways of choosing r elements out of n, order matters (number of permutations)

Let P_n^r be the number of ways of choosing r elements out of n, when the order matters. Then

$$P_n^r=\frac{n!}{(n-r)!}. \tag{2.14}$$

Case C: no repetition, order doesn't matter (Combinations)

If we want to get r successive choices from a set of n elements, not allowing repetition and with order not mattering, the number of outcomes is named as C_n^r, the C referring to the fact that choices of this kind are called *combinations* of r elements from n.

Theorem 2.5 **Number of ways of choosing r elements out of n, order does not matter (number of combinations)**

Let C_n^r be the number of ways of choosing r elements out of n, when the order does not matter. Then

$$C_n^r = \frac{n!}{r!(n-r)!}. \tag{2.15}$$

We can always write

$$C_n^r = \frac{n(n-1)\cdots(n-r+1)}{r!}.$$

Particularly

$$C_n^2 = \frac{n(n-1)}{2},$$

$$C_n^3 = \frac{n(n-1)(n-2)}{6},$$

and so on.

【Example 2.7】 (National Lottery) On single ticket, you must choose 6 integers between 1 and 49. A machine also selects 6 integers between 1 and 49, and it is reasonable to assume that each number is equally likely to be chosen.

We can write the sample space as

$S = \{(x_1, x_2, x_3, x_4, x_5, x_6): x_i \in \{1, 2, \cdots, 49\}$ for $i = 1, 2, \cdots 6$ and $x_i \neq x_j$ for $i \neq j\}$.

If I buy one ticket, what is the probability that I match all 6 numbers?

Solution: We assume the event $A = \{$one ticket matches all 6 numbers$\}$.

$$P(A) = \frac{P_6^6}{C_{49}^6} = \frac{1}{P_{49}^6} = \frac{43!}{49!}.$$

2.4.3 Estimated Probability[1] (Relative Frequency)

The frequency of the event is the number of times that the event occurs. The relative frequency or estimated probability of the event is the fraction of times the event occurs. We assume N is the number of times that the experiment is performed. We also called N the sample size. The number of times the event E occurs is $f_r(E)$, then the estimated probability $P(E)$ of the event is

$$P(E) = \frac{f_r(E)}{N}. \tag{2.16}$$

The idea is that the more frequently an event occurs, the higher the probability of that event is. We can assign a probability of an event A as follows.

---

[1] estimated probability：估计概率

(1) Repeat the experiment a large number of times.

(2) Set $P(E)$ to be the proportion of experiments in which E occurs.

In one football match, Team A has played with Team B 300 times, with 100 wins, 150 losses, and 50 draws. We could then state

$$P(\text{Team A win})=\frac{100}{300},$$

$$P(\text{Team B win})=\frac{150}{300},$$

$$P(\text{draw})=\frac{50}{300}.$$

There are also problems with this approach. Firstly, we haven't actually repeated the *same* experiment 300 times, as three matches were played for long period, and the relative strengths of the two teams will have changed over that period. When using the frequency approach, we need to think carefully about we really are repeating the same experiment lots of times. The second problem with the frequency approach is that even if we can repeat the same experiment lots of times, the proportion of times the event occurs will change after each experiment. To deal with this, a probability based the frequency approach is defined to be the limiting value of the proportion, as the number of trials tends to infinity (We will discuss this further in the study of the "law of large numbers"). This doesn't help us *assign* probabilities in practice, but it does give an *interpretation* of what a probability means.

2.4.4 Subjective Probabilities[1]

Subjective probability is a type of probability derived from an individual's personal judgment or own experience about whether a specific outcome is likely to occur. It contains no formal calculations and only reflects the subject's opinions and past experience. Subjective probabilities differ from person to person and contain a high degree of personal bias. In the subjective approach, there is no such thing as *the probability* of an event. Instead, an individual chooses a probability to represent his or her "degree of belief" on a scale of 0, meaning "I am certain that A will not occur", to 1, meaning "I am certain that A will occur". Two different people may have different probabilities for the same event, depending on their knowledge and beliefs. Subjective probability can be affected by a variety of personal beliefs held by an individual. These could relate back to upbringing as well as other events the person has witnessed throughout his life. Even if the individual's

[1] subjective Probabilities：主观概率

belief can be rationally explained, it does not make the prediction an actual fact. It is often based on how each individual interprets the information presented to him.

2.5 Conditional Probability[1]

Conditional probability is an important concept that we can use to change a measurement of uncertainty as our information changes.

【Example 2.8】 For a randomly selected individual, suppose the probabilities of the four blood types are $P(\text{type O})=0.45$, $P(\text{type A})=0.4$, $P(\text{type B})=0.1$ and $P(\text{type AB})=0.05$. A test is taken to determine the blood type, but the test is only able to declare that the blood type is either A or B. What is the probability that the blood type is A?

Definition 2.17 Conditional probability

We define $P(E|F)$ to be the **conditional probability of E given F** (see Figure 2.6), where

$$P(E|F)=\frac{P(E\cap F)}{P(F)}. \tag{2.17}$$

Assuming $P(F)>0$.

We can interpret this to mean "If it is known that F has occurred, what is the probability that E has also occurred?" If we know that the outcome belongs to the set F, then for E to occur too, the outcome should set in the intersection $E\cap F$. To get the conditional probability of $E \mid F$, we 'measure' (using the probability measure P) the fraction of the set F that is also in the set E.

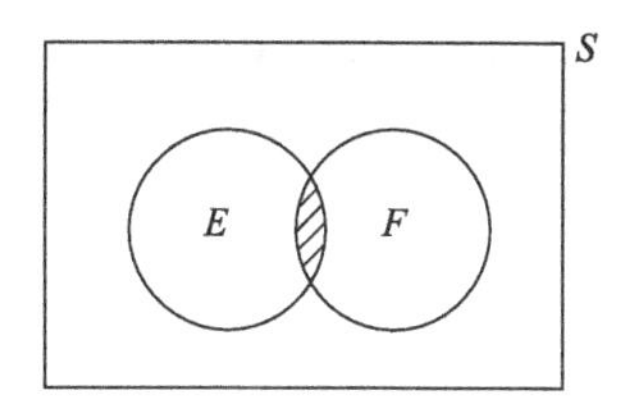

Figure 2.6 $P(E|F)$

We know that the outcome is a member of the set F, and want the probability that the outcome is also in E. To understand this result using classical probability, suppose we have a population of 100 people, in which 45 are type O, 40 are type A, 10 are type B, and 5 are type AB. If we select one person at random, then $P(\text{type O})=45/100$, $P(\text{type A})=40/100$, $P(\text{type B})=10/100$ and $P(\text{type AB})=5/100$. Now suppose we are told that the person we have selected is either type A or type B. That leaves us with 50 people that the person could be. Out of those 50, 40 are type A, and, given that each person out of the 50 had the same chance

[1] conditional probability：条件概率

of being selected,

$$P(\text{type A given that the person is either type A or type B})=0.8.$$

Note that we can write

$$\frac{40}{50}=\frac{40/100}{50/100}=\frac{P(\text{type A})}{P(\text{type A or type B})}=\frac{P(E\cap F)}{P(F)}.$$

2.5.1 Independence[1]

Definition 2.18 Independence of two events

Two events E and F are said to be **independent** if

$$P(E\cap F)=P(E)P(F). \tag{2.18}$$

We can use the definition of conditional probability to give a more intuitive definition of independence. The events E and F are independent if

$$P(E|F)=P(E). \tag{2.19}$$

We can read this to mean "If E and F are independent, then learning that E has occurred does not change the probability that F will occur (and vice versa)." (If (2.19) holds, then $P(E\cap F)=P(E|F)P(F)=P(E)P(F)$).

【Example 2.9】 A company has five factories manufacturing computers. In one factory, 2% of the computers have been built with faulty hard disks, but it is not known in which of the three factories this has happened. The hard disks in the other two factories are not faulty. You buy one computer. Let E be the event that the computer came from the 'bad' factory and F be the event that the computer has a faulty hard disk. Given the information, we might suppose that

$$P(E)=\frac{1}{5},$$

$$P(F|E)=\frac{2}{100},$$

$$P(F)=\frac{1}{3}\times\frac{2}{100}\neq P(F|E),$$

so that F and E are not independent.

Now suppose that you buy a second computer, with G the event that this computer has a faulty hard disk. If we have no information regarding whether the two computers came from the same factory, then we may judge G and F to be independent:

$$P(G|F)=P(G)=\frac{1}{3}\times\frac{2}{100}.$$

Assuming that learning that F has occurred tells us nothing about which factory the second computer came from, the probability of G would be unchanged. But if we

[1] independence：独立性

also knew that the two computers came from the same factory, G and F would no longer be independent. Learning that F has occurred would eliminate the possibility that the two computers came from a 'good' factory, and

$$P(G|F)=\frac{2}{100}.$$

2.5.2 The Law of Total Probability❶

In some situations, calculating a probability of an event is easiest if we first consider some appropriate conditional probabilities.

【Example 2.10】 Suppose the four teams in this year's Champions League semi-finals are Manchester United, Barcelona, Milan, and Bayern Munich. Depending on their opponents, you judge Manchester United's probabilities of reaching the final to be

P(Manchester United reaches final | opponent is Barcelona) $=0.2$,

P(Manchester United reaches final | opponent is Milan) $=0.4$,

P(Manchester United reaches final | opponent is Munich) $=0.6$.

If the semi-final draw has yet to be made (with any two teams having the same probability of being drawn against each other), what is your probability of Manchester United reaching the final?

If general, suppose we have a partition of $\varepsilon=\{E_1,E_2,\cdots,E_n\}$ of a sample space S. Then for any event F,

$$\begin{aligned} F &= F\cap S \\ &= F\cap(E_1\cup E_2\cup\cdots E_n) \ (\varepsilon \text{ is a partition}) \\ &= (F\cap E_1)\cup(F\cap E_2)\cup\cdots\cup(F\cap E_n) \ (\text{distributivity}) \end{aligned} \tag{2.20}$$

and since $(F\cap E_i)\cup(F\cap E_j)=\phi$ for any $i\neq j$,

$$P(F)=\sum_{i=1}^{n}P(F\cap E_i), \tag{2.21}$$

or, equivalently,

$$P(F)=\sum_{i=1}^{n}P(F|E_i)P(E_i). \tag{2.22}$$

These two equations are known as the **law of total probability.**

Given a partition $\varepsilon=\{E_1,E_2,\cdots,E_n\}$, we can calculate $P(F)$ by considering in turn how likely each event E_i is, and then how likely F is conditional on E_i. Figure 2.7 shows an example of $n=5$.

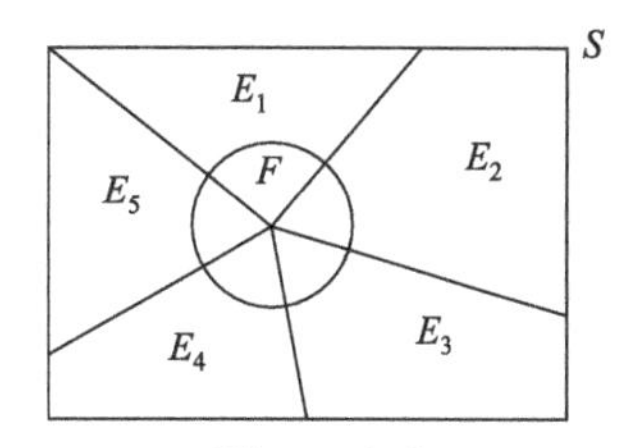

Figure 2.7

An example of $n=5$

❶ total probability：全概率

In Example 2.9, F was the event of Manchester United reaching the final, and the partition ε was the three possible opponents that they could have been drawn against.

$F=\{\text{Manchester United reaches final}\}$,

$E_1=\{\text{Manchester United's opponent is Barcelona}\}$,

$E_2=\{\text{Manchester United's opponent is Milan}\}$,

$E_3=\{\text{Manchester United's opponent is Munich}\}$.

$$\begin{aligned}P(F)&=P(F\cap(E_1\cup E_2\cup E_3))=P(F\cap E_1)+P(F\cap E_2)+P(F\cap E_3)\\&=P(E_1)P(F|E_1)+P(E_1)P(F|E_1)+P(E_1)P(F|E_1)\\&=\frac{1}{3}\times(0.2+0.4+0.6)=0.4.\end{aligned}$$

So, the probability of Manchester United reaches final is 0.4.

2.5.3 Bayes' Theorem[1]

【Example 2.11】 A new diagnostic test has been developed for a particular disease. It is known that 0.1% of people in the population have the disease. The test will detect the disease in 95% of all people who really do have the disease. However, there is also the possibility of a "false positive"; out of all people who do not have the disease, the test will claim they do in 2% of cases.

A person is chosen at random to take the test, and the result is "positive". How likely is it that that person has the disease?

Let E be the event that the person has the disease, and F be the event that the test result is positive. We have been given the information

$$P(E)=0.001, P(F|E)=0.95, P(F|\overline{E})=0.02.$$

We want $P(E|F)$: the probability of having the disease, given a positive test result. How do we get it from the probabilities that we have been told?

From the definition of conditional probability, we have

$$P(E|F)=\frac{P(E\cap F)}{P(F)}, \tag{2.23}$$

$$P(F|E)=\frac{P(F\cap E)}{P(E)}. \tag{2.24}$$

Therefore

$$P(E\cap F)=P(E|F)P(F)=P(F|E)P(E). \tag{2.25}$$

We now have the result

$$P(E|F)=\frac{P(F|E)P(E)}{P(F)}, \tag{2.26}$$

[1] Bayes' theorem：贝叶斯定理

which is known as **Bayes' theorem.** But what is about $P(F)$? Using the law of total probability, noticing that $\{E, \bar{E}\}$ is a partition, we have

$$P(F)=P(F|E)P(E)+P(F|\bar{E})P(\bar{E}). \tag{2.27}$$

Hence Bayes' theorem can be written as

$$P(E|F)=\frac{P(F|E)P(E)}{P(F|E)P(E)+P(F|\bar{E})P(\bar{E})}. \tag{2.28}$$

More generally, we state **Bayes' theorem** as

$$P(E_i \mid F)=\frac{P(E_i)P(F \mid E_i)}{\sum_{j=1}^{n} P(E_i)P(F \mid E_i)} \tag{2.29}$$

for a partition $\{E_1, E_2, \cdots, E_n\}$.

Returning to the example 2.10,

$$P(E|F)=\frac{0.95\times 0.001}{0.95\times 0.001+0.999\times 0.02}=0.045.$$

The probability of having the disease, given the positive test result, is 0.045, which may seem rather small. Remember, however, that *before* taking the test, the probability was 0.001, so the chances of having the disease have gone up from 1 in 1000, to about 1 in 20.

In the context of Bayes' theorem, we sometimes refer to $P(E)$ as the **prior probability**[1] of E, and $P(E|F)$ as the **posterior probability**[2] of E given F. The prior probability states how likely we thought E was *before* we knew that F had occurred, and the posterior probability states how likely we think E is *after* we have learnt that F has occurred.

Exercises

1. Assume the events A, B, C, please express the following event:
   (1) A happens, B, C do not happen;
   (2) A, B happen, C does not happen;
   (3) A, B, C all happen;
   (4) A, B, C all do not happen;
   (5) At least one happens among A, B, C;
   (6) At least two happen among A, B, C.

---

[1] prior probability：先验概率
[2] posterior probability：后验概率

2. Assume $P(A)=P(B)=P(C)=\frac{1}{4}$, $P(AB)=P(BC)=0$, $P(AC)=\frac{1}{8}$, please compute the probability of one event happens at least among the three events.

3. Assume $P(A)=\frac{1}{2}$, $P(B)=\frac{1}{3}$, $P(C)=\frac{1}{5}$, $P(AB)=\frac{1}{10}$, $P(BC)=\frac{1}{20}$, $P(AC)=\frac{1}{15}$, $P(ABC)=\frac{1}{30}$, please compute $P(A\cup B)$, $P(\overline{AB})$, $P(A\cup B\cup C)$, $P(\overline{A}\,\overline{B}\,\overline{C})$, $P(\overline{A}\,\overline{B}C)$, $P(\overline{A}\,\overline{B}\cup C)$.

4. If $S=\{s,m,d\}$, then the power set $P(S)=\{\phi,\{s\},\{m\},\{d\},\{s,m\},\{s,d\},\{m,d\},S\}$ is a Boolean algebra. Why is

$$G=\{\{s\},\{m\},\{d\}\}$$

not a Boolean algebra?

5. For any three sets $A,B,C\in B(S)$ (not necessarily disjoint), prove that for any measure m,

$$m(A\cup B\cup C)=m(A)+m(B)+m(C)-m(A\cap B)-m(A\cap C)-m(B\cap C)+m(A\cap B\cap C)$$

You may quote the result that $A\cap(B\cup C)=(A\cap B)\cup(A\cap C)$.

6. (1) $S=\{a,b,c\}$. Write down two partitions of S, and one collection of sets that is not a partition.

(2) $S=R^{+}$. Write down two partitions of S.

7. Prove: $P(\overline{A})=1-P(A)$.

8. There are 10 students in one room. They wear the badges from 1 to 10 respectively. We select three students at random and write down the numbers of theirs badges.

Please answer the following questions.

(1) The probability of the minimum number is 5;

(2) The probability of the maximum number is 5.

9. In the National Lottery, what is the probability of matching exactly 3 numbers?

10. If we are given

$$P(E)=\frac{1}{3},\ P(F\mid E)=\frac{2}{100},$$

How do we get $P(F\cap E)$?

11. (1) Assume $P(A)=\frac{1}{2}$, $P(B)=\frac{1}{4}$, $P(B\mid A)=\frac{1}{3}$, $P(A\mid B)=\frac{1}{2}$, please compute $P(A\cup B)$.

(2) Assume $P(\overline{A})=0.3$, $P(B)=0.4$, $P(A\overline{B})=0.5$, please compute

$P(B|A\cup\overline{B})$.

12. Please prove:

(1) If $P(A)>0$, then $P(AB|A)\geqslant P(AB|A\cup B)$;

(2) If $P(A|B)=1$, then $P(\overline{B}|\overline{A})=1$;

(3) If $P(A|C)\geqslant P(B|C)$, $P(A|\overline{C})\geqslant P(B|\overline{C})$, then $P(A)\geqslant P(B)$.

13. There are 10 goods including 2 defective goods. We select one in 2 defective goods with non-return sample. Please compute the following probability.

(1) The selected two goods are defective;

(2) The selected two goods are normal;

(3) The selected one good is defective, the other one is normal;

(4) The second selected one is defective.

14. Please prove the following:

(1) If A, B, C are mutual independent, then C and AB are independent;

(2) If $P(A)=0$, then A and B are independent for arbitrary event B.

15. There are the same goods in two boxes. There are 50 goods in the first box including 10 normal goods and there are 30 goods in the second box including 18 normal goods. We select one from two boxes randomly and select goods two times and select one each time with non-return sample. Please compute the following probabilities.

(1) The goods are normal at the first time;

(2) The second is also normal in the condition of the first is normal.

Chapter 3 Discrete Random Variables

In many experiments, the outcomes are of a numerical nature, e. g. , if they correspond to stock prices or the number of points that appear on the dice. In other experiments, the outcomes are not numerical, but they may be associated with some numerical values of interest. For example, if the experiment is to see whether a coin toss appears head or tail, we can associate head to number 1 and tail to number 0. When dealing with such numerical values, it is often useful to assign probabilities to them. This is done through the notion of a random variable. There are two fundamentally different types of random variables—discrete random variables and continuous random variables. In this chapter, our focus is on discrete random variables. We discuss the basic properties and the most important examples of one-dimensional and multidimensional random variables.

3.1 Random Variables

Informally, we think of a random variable as any quantity that is uncertain to us. For example:

- The number of broken machines in a workshop during a certain hour;
- The number of times a team wins in 10 football matches.

We cannot say with certainty what any of these quantities are, but probability theory gives us a framework for describing how likely different values are. Whereas elements of a sample space may not be numerical, random variables are always numerical quantities, and so, when defining a random variable, we need a rule for getting from the random outcome in the sample space to the value of the random variable. Such a rule of association is called a random variable—a variable because different numerical values are possible and random because the observed value depends on which of the possible experimental outcomes results .

Definition 3.1 (**Random variable**) Given a sample space S, we define a random variable (r. v) X to be a mapping from S to the real line R. We write a random variable as $x = X(\omega)$, where $\omega \in S$.

Random variables are customarily denoted by uppercase letters, such as X and Y. Lowercase letters are used to represent some particular value of the corresponding random variable. The notation $X(\omega) = x$ means that x is the value associated with the outcome ω by the r. v X.

Remark:

- Random variables are different from ordinary functions, the latter are defined on the real number axis, while random variables are defined on the sample space.
- The value of random variable has a certain probability law. The appearance of each result of the experiment has a certain probability, therefore, the value of random variables also has a certain probability law.

【Example 3.1】 The possible results of tossing a coin are $H = \{\text{Head}\}$ and $T = \{\text{Tail}\}$, With $S = \{H, T\}$, as shown in Figure 3.1 define a r. v X by $X(H) = 1, X(T) = 0$.

The r. v X in Example 3.1 was specified by explicitly listing each element of S and the associated number. Such a listing is tedious if S contains more than a few outcomes, but it can frequently be avoided.

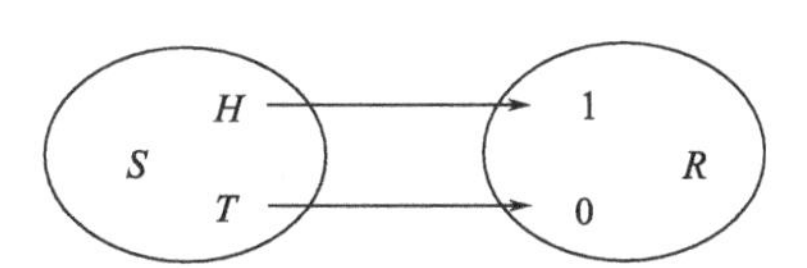

Figure 3.1 A random variable

A distinction characterizes two different types of random variables, discrete and continuous variable.

Definition 3.2 (**Discrete Random Variable & Continuous Random Variable**)

A **discrete random variable**[1] is an r. v whose possible values either constitute a finite set or else can be listed in a "countably" infinite sequence;

A **continuous random variable**[2] is an r. v whose possible values can be continuously filled with an interval.

【Example 3.2】 Let a shooter shoot the target 30 times in a row, with X as the number of times the target has been hit, then possible values of X are $D = \{0, 1, 2, \cdots, 30\}$. Since the possible values have been listed in a finite sequence, X is a discrete r. v; If the random variable Y is recorded as the number of shots fired continuously until it hits the target, the possible values of Y are $D = \{1, 2, \cdots\}$. Since

---

[1] discrete random variable：离散型随机变量

[2] continuous random variable：连续型随机变量

the possible values have been listed in a "countably" infinite sequence, Y is also a discrete r. v. If the random variable Z represents the lifetime of a bulb, the possible values of Z are filled with the interval $[0, +\infty)$, therefore Z is a continuous random.

3.2 Probability Distributions for Discrete Random Variables

In order to understand the probability law of the value of random variables, we need to study the range of possible values of random variables and the probability of taking these values, that is, the **probability distribution** of random variables. The understanding of random variables should start with its probability distribution.

To study basic properties of discrete r. v's, only the tools of discrete mathematics— summation and differences—are required. The study of continuous variables requires the continuous mathematics of the calculus—integrals and derivatives. In this chapter, we focus exclusively on discrete random variables.

3.2.1 Probability Mass Function (PMF)

Definition 3.3 The **probability distribution**[1] or **probability mass function (PMF)**[2] of a discrete r. v is defined for every number x by $p_X(x)=P\{X=x\}$. The probability distribution of discrete random variables can also be expressed in as the following matrix form or table form, respectively:

$$X \sim \begin{pmatrix} x_1 & x_2 & \cdots & x_n & \cdots \\ p_1 & p_2 & \cdots & p_n & \cdots \end{pmatrix}$$

or

X	x_1	x_2	$\cdots$	x_n	$\cdots$
p	p_1	p_2	$\cdots$	p_n	$\cdots$

Proposition 3.1 A probability mass function (PMF) must have the following two properties:

1. $p_X(x) \geqslant 0$;
2. $\sum_x p_X(x) = 1$.

---

[1] probability distribution：概率分布

[2] probability mass function (PMF)：概率质量函数

In the summation above, x ranges over all the possible numerical values of X. This follows from the additivity and normalization axioms, because the events $\{X=x\}$ are disjoint and form a partition of the sample space, as x ranges over all possible values of X.

By a similar argument, for any set S of real numbers, we also have

$$P\{X \in S\} = \sum_{x \in S} p_X(x).$$

【Example 3.3】 There are four defective products in ten products, of which two are randomly selected. Take X as the number of defective products contained in the two selected products. Obviously, the possible values of X are 0, 1, 2, and the probability that X takes i is

$$P\{X=i\}=\frac{C_4^i C_6^{2-i}}{C_{10}^2} \quad (i=0,1,2)$$

which constitutes PMF of X. Based on this, the probability of X taking a certain range of values can be calculated. For example,

$$P\{0 \leqslant X<2\}=P\{X=0\}+P\{X=1\}=0.8666;$$

$$P\{X>2\}=P(\phi)=0;$$

$$P\{X \leqslant 2\}=P(\Omega)=P\{X=0\}+P\{X=1\}+P\{X=2\}=1.$$

As seen from the Example 3.3, we can find out the probability of random variables in any range, as long as we know the PMF of the discrete random variable,

【Example 3.4】 Suppose the PMF of X is as follows:

$$X \sim \begin{pmatrix} -1 & 0 & 1 & 2 \\ \frac{1}{2c} & \frac{3}{4c} & \frac{5}{8c} & \frac{7}{16c} \end{pmatrix}$$

Based on normalization, we have $\frac{1}{2c}+\frac{3}{4c}+\frac{5}{8c}+\frac{7}{16c}=1$ to get $c=2.3125$.

We can even obtain the conditional probability based on PMF:

$$P\{X<1 \mid X \neq 0\}=\frac{P\{X<1, X \neq 0\}}{P\{X \neq 0\}}=\frac{P\{X=-1\}}{1-P\{X=0\}}=0.32.$$

3.2.2 Cumulative Distribution Function (CDF)

Although PMF intuitively gives the probability that a random variable takes any possible value, sometimes we are more concerned about the probability that the random variable will be at most a certain value, for example, the probability that a venture investor will suffer a loss up to \$10,000, the probability that a restaurant will receive at most 100 customers over a period of time. That is to say, for any fixed value x, if we are told the probability that the observed value of X will be at most x. it will be convenient in these cases. Such a tool is **cumulative dis-**

tribution function.

Definition 3.4 We define the **cumulative distribution function (CDF)**[1] $F_X(x)$ of a random variable X (discrete or continuous) to be

$$F_X(x)=P\{X\leqslant x\}. \tag{3.1}$$

From the definition, we obtain the following properties of CDF:

Proposition 3.2 A cumulative distribution function (CDF) $F_X(x)$ must have the following properties:

1. $F_X(x)$ is a monotone non-decreasing function and $0\leqslant F_X(x)\leqslant 1$; (3.2)
2. $F_X(+\infty)=\lim\limits_{x\to+\infty} F_X(x)=1, F_X(-\infty)=\lim\limits_{x\to-\infty} F_X(x)=0.$ (3.3)

Proof: Obviously, $0\leqslant F_X(x)\leqslant 1$ because $F_X(x)$ is defined as probability.
For any $x_1<x_2$, let $A=\{X\leqslant x_1\}, B=\{X\leqslant x_2\}$, then $A\subset B$, so $P(A)\leqslant P(B)$, i. e. $F_X(x_1)\leqslant F_X(x_1)$, therefore $F_X(x)$ is a non-decreasing function.
Because of boundedness and monotonicity, we have

$$F_X(+\infty)=\lim_{x\to+\infty} F_X(x)=1, F_X(-\infty)=\lim_{x\to-\infty} F_X(x)=0.$$

Note:

- It is worth to note that the probability of X lying in some interval between a and b depends on whether the lower limit a or the upper limit b is included in the probability calculation:

$P\{a<X\leqslant b\}=P\{X\leqslant b\}-P\{X\leqslant a\}=F(b)-F(a)$;
$P\{a\leqslant X\leqslant b\}=P\{a<X\leqslant b\}+P\{X=a\}=F(b)-F(a)+P\{X=a\}$;
$P\{a<X<b\}=P\{a<X\leqslant b\}-P\{X=b\}=F(b)-F(a)-P\{X=b\}$;
$P\{a\leqslant X<b\}=P\{a<X\leqslant b\}-P\{X=b\}+P\{X=a\}=F(b)-F(a)-P\{X=b\}+P\{X=a\}$.

【Example 3.5】 Let $F_X(x)=A+B\arctan x\,(x\in R)$ be the CDF of random variable X.

(1) Determine the coefficients A, B;

(2) Calculate the probability of $P\{-1<X\leqslant\sqrt{3}\}$.

Solution: (1) According to the limit property of CDF, we have

$$\begin{cases} 0=F(-\infty)=\lim\limits_{x\to-\infty} F(x)=A-\dfrac{\pi}{2}B \\ 1=F(+\infty)=\lim\limits_{x\to+\infty} F(x)=A+\dfrac{\pi}{2}B \end{cases}$$

to get $A=\dfrac{1}{2}$, $B=\dfrac{1}{\pi}$.

---

[1] cumulative distribution function (CDF)：累积分布函数

(2) The probability of random variable X in any interval can be calculated by CDF.

$$P\{-1<X\leqslant\sqrt{3}\}=F(\sqrt{3})-F(-1)=\frac{1}{\pi}\times\frac{\pi}{3}-\frac{1}{\pi}\times\left(-\frac{\pi}{4}\right)=\frac{7}{12}.$$

It can be seen from above that the probability of a discrete random variables in any range can be calculated according to either CDF or PMF, both of which are the distribution characteristics of random variables. In fact, the two are equivalent and can be deduced mutually.

Proposition 3.3 (Mutual deduction between CDF and PMF)

The cumulative distribution function (CDF) can be written in terms of the probability mass function (PMF):

$$F_X(x)=P(X\leqslant x)=\sum_{a\leqslant x}p_X(a).$$

The probability mass function (PMF) can be written in terms of the cumulative distribution function (CDF):

$$P_X(x_i)=F_X(x_i)-F_X(x_{i-1}),i>1.$$

【Example 3.6】 Suppose that the PMF of r. v X is $X\sim\begin{pmatrix}-1 & 2 & 3\\ \frac{1}{4} & \frac{1}{2} & \frac{1}{4}\end{pmatrix}$.

(1) Find the CDF of X;

(2) Can PMF of X be calculated according to the expression of CDF of X?

Solution:

(1) According to the definition and the interval where x is located (see Figure 3.2), we obtained the CDF as follows (see Figure 3.3):

$$F_X(x)=P\{X\leqslant x\}=\begin{cases}P(\phi), & x<-1,\\ P\{X=-1\}, & -1\leqslant x<2,\\ P\{X=-1\}+P\{X=2\}, & 2\leqslant x<3,\\ P(\Omega), & x\geqslant 3\end{cases}=\begin{cases}0, & x<-1,\\ \frac{1}{4}, & -1\leqslant x<2,\\ \frac{3}{4}, & 2\leqslant x<3,\\ 1, & x\geqslant 3.\end{cases} \quad (3.4)$$

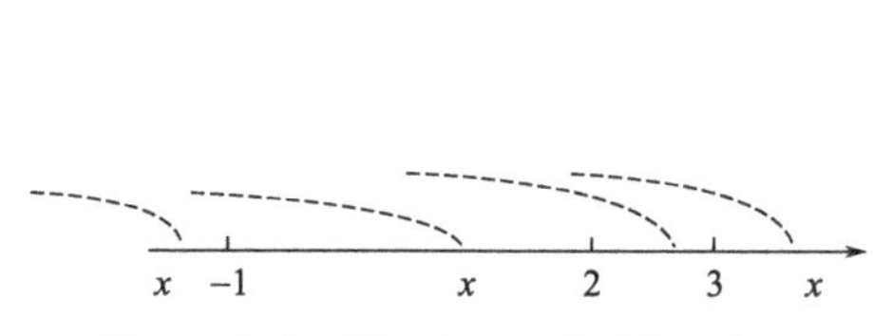

Figure 3.2 The interval of Random variable X based on x

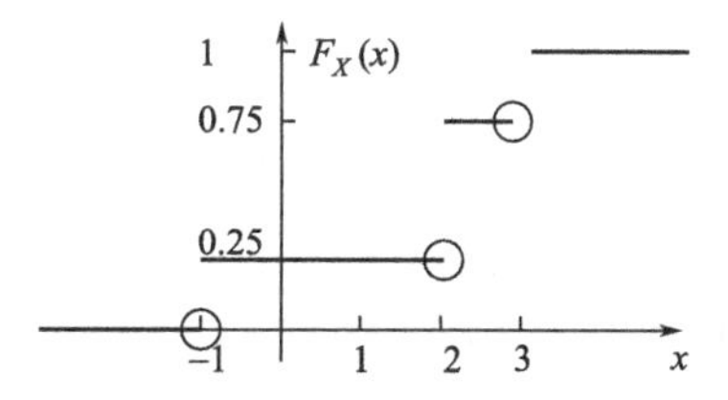

Figure 3.3 Cumulative distribution function diagram

As can be seen from Figure 3. 3 and equation (3. 4), the CDF of the discrete random variable is a monotonic piecewise function, and its segmented point is the possible value of the random variable. The image of CDF rises step by step and is right continuous at the piecewise point, continuous at other points.

(2) From the CDF of X, we know that the possible values of X are $-1,2,3$, the corresponding probabilities are as follows:

$$P\{X=-1\}=P\{X\leqslant -1\}-P\{X<-1\}=F(-1)-F(x)=\frac{1}{4}-0=\frac{1}{4}\ (x<-1);$$

$$P\{X=2\}=P\{X\leqslant 2\}-P\{X\leqslant -1\}=F(2)-F(-1)=\frac{3}{4}-\frac{1}{4}=\frac{1}{2};$$

$$P\{X=3\}=P\{X\leqslant 3\}-P\{X\leqslant 2\}=F(3)-F(2)=1-\frac{3}{4}=\frac{1}{4};$$

Which constitute the PMF of X.

Note:

- Both PMF and CDF can uniquely characterize the distribution characteristics of discrete random variables, but for discrete random variables, PMF is more intuitive, so it is generally used to analyze the probability distribution of discrete random variables.

3. 2. 3 Derived Distributions of Discrete Random Variables

Consider the sales revenue of a company, let the random variable X be the sales volume and the unit price of the product is 3, then the sales revenue $Y=3X$. In this example, Y is the function of X, In general, if $Y=g(X)$ is a function of a random variable X, then Y is also a random variable, since it provides a numerical value for each possible outcome. This is because every outcome in the sample space defines a numerical value x for X and hence also the numerical value $y=g(x)$ for Y. If X is discrete with PMF p_X, then Y is also discrete, and its PMF p_Y can be calculated using the PMF of X. In particular, to obtain $p_Y(y)$ for any y, we add the probabilities of all values of x such that $g(x)=y$,

$$p_Y(y)=\sum_{\{x\mid g(x)=y\}} p_X(x). \tag{3.5}$$

【Example 3. 7】 Let Y be the function of discrete random variable X and the PMF of X is as follows:

$$X\sim\begin{pmatrix} -1 & 0 & 1 & 2 & 3 \\ 0.2 & 0.2 & 0.1 & 0.3 & 0.2 \end{pmatrix}.$$

Try to find the PMF of Y when $Y=X+1$ or $Y=(X-1)^2$.

Solution: (1) If $Y=X+1$, the possible values of Y are $0,1,2,3,4$, and

$p_Y(0)=p_X(-1)=0.2$; $p_Y(1)=p_X(0)=0.2$; $p_Y(2)=p_X(1)=0.1$;

$p_Y(3)=p_X(2)=0.3$; $p_Y(4)=p_X(3)=0.2$.

Tabulate the PMF,

$$Y \sim \begin{pmatrix} 0 & 1 & 2 & 3 & 4 \\ 0.2 & 0.2 & 0.1 & 0.3 & 0.2 \end{pmatrix}.$$

(2) If $Y=(X-1)^2$, the possible values of Y are 0, 1, 4 and

$$p_Y(0)=p_X(1)=0.1;$$
$$p_Y(1)=p_X(0)+p_X(2)=0.2+0.3=0.5;$$
$$p_Y(4)=p_X(-1)+p_X(3)=0.2+0.2=0.4.$$

i. e.

$$Y \sim \begin{pmatrix} 0 & 1 & 4 \\ 0.1 & 0.5 & 0.4 \end{pmatrix}.$$

3.3 Some Important Discrete Probability Distributions

We now consider some standard probability distributions for discrete random variables, which can be used in a variety of different applications. By "distribution", we mean a particular choice of probability mass function, which may be specified in terms of some parameters.

3.3.1 The Bernoulli Distribution

We often face situations where there are only two possible outcomes. For example, a football team wins or loses a match, tossing an unbiased coin comes up a head or tail, the SP500 index rises or falls. All these results can be described by Bernoulli random variable. A Bernoulli random variable X can take one of two values: 0 and 1.

Definition 3.5 (**Bernoulli distribution**[1]) If a random variable X has a Bernoulli distribution with parameter p, then its PMF is

$$X \sim \begin{pmatrix} 1 & 0 \\ p & 1-p \end{pmatrix},$$

or

$$p_X(x)=p^x(1-p)^{1-x}, \tag{3.6}$$

where $x=0,1; 0<p<1$, we write

$$X \sim \text{Bernoulli}(p).$$

【Example 3.8】 If X denotes the number of times a uniform coin comes up a head

[1] Bernoulli distribution：伯努利分布

when it is tossed once. The results are head or tail. Thus X can only be 1 or 0. So $X \sim \text{Bernoulli}(0.5)$. At the same time, if Y denotes the number of times this coin comes up a tail when it is tossed once. The results are also head or tail. Thus Y can only be 0 or 1. So Y also follows Bernoulli distribution, $Y \sim \text{Bernoulli}(0.5)$, though they are different random variables.

Note:

- Different random variables can have exactly the same probability distribution. Such as, whether the newborn baby is male or female, whether it rains tomorrow, whether the seeds germinate etc, all of the random phenomenon can be modeled by Bernoulli distribution.

3.3.2 The Binomial Distribution

Now let's consider repeating the above experiment with only two outcomes n times. We have a sequence of n trials, and in each trial we observe a 'success' or a 'failure'. The probability of a success within each trial, p, is the same, and the trials are independent. Take X as the total number of successes, what is the probability distribution of X? Examples of 'experiments' that we might describe using this distribution are, the number of lotteries win in 100 lotteries, the number of times a tennis player wins in five games, the number of times a student scores more than 90 points in eight exams.

Definition 3.6 (**Binomial distribution**[1]) If a random variable X has a binomial distribution, with parameters n (the number of trials) and p (the probability of success in each trial), then the PMF of X is

$$p_X(x) = C_n^x p^x (1-p)^{n-x}, \tag{3.7}$$

where $0 < p < 1, x = 0, 1, 2, \cdots, n$, we write

$$X \sim \text{Bin}(n, p).$$

To understand the PMF of X, we suppose that all the results of an experiment satisfy the condition that event A occurs with probability p and does not occur with probability $1-p$; The experiment is carried out independently and repeatedly n times, suppose B_i indicates that event A occurred in the ith experiment, X is the number of times event A occurred in these n experiments, and event B means that event A occurs x times in n independent repeated experiments, we consider the probability of event B.

Obviously, $P(B_i) = P(A) = p, P(\overline{B_i}) = P(\overline{A}) = 1-p, i = 1, 2, \cdots n.$

In n experiments, as long as event A happens to occur x times, event B oc-

---

[1] Binomial distribution：二项分布

curs, regardless of the order in which these As occur, so event B occurs means that at least one of the C_n^p results occurs,

$$B=B_1B_2\cdots B_k\overline{B}_{k+1}\cdots\overline{B}_n+B_1\cdots B_{k-1}\overline{B}_kB_{k+1}\overline{B}_{k+2}\cdots\overline{B}_n+\cdots+\overline{B}_1\cdots\overline{B}_{n-k}B_{n-k+1}\cdots B_n,$$

and the probability of occurrence of each result is $p^x(1-p)^{n-x}$.

Thus,

$$p_X(x)=P(B)=C_n^xp^x(1-p)^{n-x}.$$

Note that the binomial PMF is valid (i. e it sums to 1). Based on the binomial theorem, we have

$$\sum_{x=0}^{n}p_X(x)=\sum_{x=0}^{n}C_n^xp^x(1-p)^{n-x}=[p+(1-p)]^n=1.$$

【Example 3.9】 In the face of five four-choose-one questions on the test paper, a student tried to give the answer by drawing lots. Try to calculate the probability of the following events:

(1) Exactly 2 questions were answered correctly;

(2) At least 2 questions were answered correctly;

(3) None of the answers are correct;

(4) All the answers are correct.

Solution: Let X be the number of questions answered correctly by the student. Because the choice of each question is independent of each other, and the probability of the correct choice of each question is 0.25, so $X\sim\text{Bin}(5,0.25)$, thus,

(1) $P\{X=2\}=C_5^2(0.25)^2(0.75)^3\approx 0.2637$;

(2) $P\{X\geqslant 2\}=\sum_{k=2}^{5}C_5^k(0.25)^k(0.75)^{5-k}\approx 0.3672$;

(3) $P\{X=0\}=(0.75)^5\approx 0.2373$;

(4) $P\{X=5\}=(0.25)^5\approx 0.0010$.

3.3.3 Hypergeometric Distributions

Hypergeometric distribution is an important distribution to describe non-return sampling, and it is one of the commonly used distributions in product inspection.

The assumptions leading to the hypergeometric distribution are as follows:

(1) The population to be sampled consists of N individuals (a finite population).

(2) Each individual can be characterized as a success (S) or a failure (F), and there are M successes in the population.

(3) A sample of n individuals is selected without replacement in such a way that each subset of size n is equal likely to be chosen.

(4) The random variable of interest is X = the number of S's in the population.

Definition 3.7 (**Hypergeometric distribution**)[1] If a random variable X has a hypergeometric distribution, with parameters n, M, and N, n, then the PMF of X is

$$p_X(x)=\frac{C_M^x C_{N-M}^{n-x}}{C_N^n}, \tag{3.8}$$

in which X is the number of S's in a completely random sample of size n drawn from a population consisting of M S's and $(N-M)$ F's, $\max(0,n-N+M)\leqslant x\leqslant\min(n,M)$, we write

$$X\sim h(n,M,N).$$

Remark:

- The Hypergeometric distribution is related to the Binomial distribution. The Binomial distribution is the approximate probability model for sampling without replacement from a finite dichotomous (S-F) population provided the sample size n is small relative to the population size N; The Hypergeometric distribution is the exact probability model for the number of S's in the sample. The Binomial r. v X is the number of S's when the number n of trials is fixed.

$$P\{X=x\}=\frac{C_M^x C_{N-M}^{n-x}}{C_N^n}\xrightarrow{\frac{n}{N}\ll 1,\frac{M}{N}=p} C_n^x p^x q^{n-x}.$$

【Example 3.10】 According to the regulations, a certain type of electronic components with a service life of more than 1500 hours is a first-class product. The first-class rate of **a large number of** products is known to be 0.2, from which 20 are now randomly selected. What is the probability of exactly k of the 20 components ($k=0,1,\cdots,20$)?

This is non-return sampling. However, because the total number of components is very large and the number of components sampled is very small relative to the total number of components, this sampling can be approximately treated as a return sampling. So, examining the 20 components is supposed to be like repeating 20 experiments, and the probability of getting the first grade at a time is 0.2. Therefore, the number of the first grade X follows the Binomial distribution,

$$X\sim \text{Bin}(20,0.2),\ \text{and}\ P(X=k)=C_{20}^k(0.2)^k(0.8)^{20-k},k=0,1,\cdots,20.$$

We get the PMF (see Figure 3.4),

$$\begin{pmatrix} 0 & 1 & 2 & 3 & 4 & 5 & 6 & 7 & 8 & 9 & 10 & \geqslant 11 \\ 0.012 & 0.058 & 0.137 & 0.205 & 0.218 & 0.175 & 0.109 & 0.055 & 0.022 & 0.007 & 0.002 & \approx 0 \end{pmatrix};$$

[1] Hypergeometric distribution：超几何分布

Remark:

• It can be seen from the Figure 3.4 that when k increases, the probability $P\{X=k\}$ firstly increases until the maximum value is taken, and then begins to decrease monotonously. Generally speaking, for fixe n and p, the probability of Binomial distribution increases at first and then decreases with the increase of the value of random variables. It will be found that a similar phenomenon occurs in Poisson distribution and normal distribution of random variables.

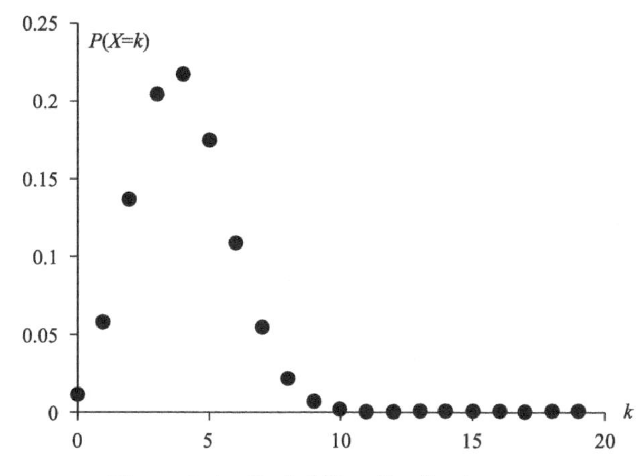

Figure 3.4 Probability distribution map

【Example 3.11】 Someone buys 40000 lottery tickets, each of which has a winning probability of 0.0001. Try to find:

(1) The probability of winning the lottery;

(2) The probability of winning at least three lottery tickets.

If the number of times to buy a lottery ticket is X, and each purchase of a lottery ticket is regarded as an experiment, then X represents the number of wins with a probability of 0.0001 in 40000 experiments, so $X \sim \text{Bin}(40000, 0.0001)$, thus,

(1) $P\{X \geqslant 1\} = 1 - P\{X=0\} = 1 - (0.9999)^{40000} \approx 0.982$.

Remark:

• The result shows that the probability of an event (such as winning the lottery) occurring in an experiment is very small (such as the winning rate here is 0.0001), but as long as the number of trials is large and carried out independently (such as buying 40000 copies here). Then the occurrence of this event (such as winning the lottery at least once) is almost certain (for example, the probability here is 0.982), that is, the event is going to happen sooner or later in the experiment.

This is commonly referred to as the **principle of small probability events**[1].

• The principle of small probability events is also commonly used on the other hand, which is called **the principle of practical inference**[2]. For example, if someone buys 40000 lottery tickets and always fails, in view of the fact that the probability of the event occurrence is very small, we have reason to suspect that the assumption that the winning rate is 0.0001 is wrong according to the actual inference principle. That is, the winning rate should be less than 0.0001.

(2) $P\{X \geqslant 3\}$

$$
\begin{aligned}
&= 1 - P\{X=0\} - P\{X=1\} - P\{X=2\} \\
&= 1 - (0.9999)^{40000} - C_{40000}^{1}(0.0001)(0.9999)^{39999} - \\
&\quad C_{40000}^{2}(0.0001)^{2}(0.9999)^{39998} \\
&\approx 0.762.
\end{aligned}
$$

3.3.4 The Poisson distribution

The Poisson distribution is used to represent **count data**: the number of times an event occurs in some finite interval in time or space. For example, the number of customers arriving at the store within an hour, the number of earthquakes in a city in a year, the number of blemishes on a piece of cloth, all of these can be modeled by Poisson distribution.

Whereas a Binomial random variable corresponds to observations of 'successes' in n trials, and so cannot be larger than the number of trials n, in theory, a Poisson random variable has no upper limit. The Poisson distribution has a single parameter λ, known as the rate parameter. We will show that λ is the average number of times the event will occur.

Definition 3.8 **(Poisson distribution)**[3] If a random variable X has a Poisson distribution, with rate parameter $\lambda > 0$, then its PMF is

$$
p_X(x) = \frac{\lambda^x e^{-\lambda}}{x!} \tag{3.9}
$$

where $x \in Z^+$ and 0 otherwise. We write

$$
X \sim \text{Poisson}(\lambda).
$$

We can confirm that this is a valid PMF according to the Taylor expansion equation of e^{λ}. In fact,

$$
\sum_{x=0}^{\infty} p_X(x) = \sum_{x=0}^{\infty} \frac{\lambda^x e^{-\lambda}}{x!} = e^{-\lambda} \sum_{x=0}^{\infty} \frac{\lambda^x}{x!} = e^{-\lambda} e^{\lambda} = 1.
$$

---

[1] principle of small probability events：小概率事件原理

[2] the principle of practical inference：实际推断原理

[3] Poisson distribution：泊松分布

It can be proved that Poisson distribution is a limit of Binomial distribution.

Let's start with a question, suppose that a shop receives μ customers in an hour on average, then, what is the probability of k customers stepping into the shop within T hours?

Let X be the number of customers stepping into the shop within T hours, divide T equally into n segments (n is large enough, that is, each period of time T/n is short enough), so it is reasonable to assume that the number of customers arriving in each period is 1 or 0, and the probability of one customer arriving in each period is $p=\mu T/n$ (as can be proved in the following sections), and the number of arrivals in each period is independent of each other. Therefore, it can be considered that X obeys Binomial distribution, that is $X\sim \text{Bin}(n,\mu T/n)$. Then,

$$\begin{aligned} P(X=k) &= C_n^k\left(\frac{\mu T}{n}\right)^k\left(1-\frac{\mu T}{n}\right)^{n-k} \\ &= \frac{n(n-1)\cdots(n-k+1)}{k!\ n^k}(\mu T)^k\left(1-\frac{\mu T}{n}\right)^{n-k} \\ &\xrightarrow{n\to\infty} \frac{(\mu T)^k}{k!}e^{-\mu T} \triangleq \frac{\lambda^k}{k!}e^{-\lambda} \end{aligned} \tag{3.10}$$

where $\lambda=\mu T=np$. So we obtain the following lemma.

Lemma 3.1 The Poisson approximation to the Binomial distribution

Consider $X\sim \text{Bin}(n,p)$, and suppose that both $n\to\infty$ and $p\to 0$, such that the product np is constant. Let $\lambda=np$. Then, as $n\to\infty$, $p_X(x)=\frac{\lambda^x}{x!}e^{-\lambda}$.

Remark:

- In any binomial experiment in which n is large and p is small, binomial distribution can be approximately replaced by Poisson distribution. $\text{Bin}(x;n,p)\approx \text{Possion}(x;\lambda)$, where $\lambda=np$. As a rule of thumb, this approximation can safely be applied if $n>50$ and $np<10$.

【Example 3.12】 Someone buys 40000 lottery tickets, each of which has a winning probability of 0.0001. Try to find the probability of winning at least three lottery tickets.

In Example 3.11, we take X as the number of times to buy 40000 lottery tickets, obviously $X\sim \text{Bin}(40000,0.0001)$ and we find that it's not easy to calculate the probability of winning at least three lottery tickets. Because $n=40000$ is big enough and $p=0.0001$ is small enough, $np=4<10$, so X approximately follows the Poisson distribution, $X\sim \text{Poisson}(4)$, so

$$P\{X\geqslant 3\}=1-P\{X=0\}-P\{X=1\}-P\{X=2\}\approx 1-e^{-4}\left(1+\frac{4}{1}+\frac{4^2}{2!}\right)\approx 0.762.$$

The probability in the above formula can also be obtained from the Poisson distribution table in table 1 in appendix.

Remark:

• Summing up the three distributions, we find that the relationship among Hypergeometric distribution, Binomial distribution and Poisson distribution is as follows:

$$P\{X=x\}=\frac{C_M^x C_{N-M}^{n-x}}{C_N^n}\xrightarrow{\frac{n}{N}<<1,\frac{M}{N}=p}C_n^x p^x(1-p)^{n-x}\xrightarrow{n>50,np=\lambda<5}\frac{\lambda^k}{k!}e^{-\lambda}.$$

3.4 Multiple Discrete Random Variables

Many problems in probability and statistics involve working simultaneously with two or more random variables. For example, X and Y represents the scores of a randomly selected student in English and mathematics, respectively. X, Y and Z denotes the results of 3 test indicators in a medical diagnosis. All of these random variables are associated with the same experiment, and their values may relate in interesting ways. This motivates us to consider probabilities involving simultaneously the numerical values of several random variables and to investigate their mutual couplings. In this section, we will extend the concepts of PMF and expectation developed so far to multiple random variables. Later on, we will also develop notions of conditioning and independence that closely parallel the ideas discussed in Chapter 2.

3.4.1 Joint Distribution

Definition 3.9 Let $X_1, X_2, \cdots, X_n$ be a random variable defined on the sample space S, then the n-dimensional vector $(X_1, X_2, \cdots, X_n)$ is called an ***n*-dimensional random variable.**

In this chapter, we mainly discuss two-dimensional random variables, and the higher-dimensional case can be similar to that of two-dimensional random variables.

Definition 3.10 Joint cumulative distribution function (Joint CDF)[1]

Let (X, Y) be a two-dimensional random variable, for any real number x, y, bivariate function $F(x, y)$ is called the **joint cumulative distribution function (Joint CDF)** of (X, Y), if

$$F(x,y)=P\{\{X\leqslant x\}\cap\{Y\leqslant y\}\}=P\{X\leqslant x, Y\leqslant y\}. \tag{3.11}$$

---

[1] Joint cumulative distribution function (Joint CDF)：联合累积分布函数

It can be seen from the Definition 3.10 that the joint cumulative distribution function is actually the probability of the intersection of two events. According to the properties of the cumulative distribution function, it is not difficult to prove that the joint cumulative distribution function has the following properties:

Proposition 3.4 Properties of two-dimensional joint cumulative distribution function $F(x,y)$:

(1) $F(x,y)$ is a monotone non-decreasing function of x and y. That is, for any fixed y, $F(x_2,y)>F(x_1,y)$ when $x_2>x_1$; For any fixed x, $F(x,y_2)>F(x,y_1)$ when $y_2>y_1$;

(2) $F(x,y)$ is right continuous with respect to x and y, that is, $F(x,y)=F(x+0,y)$, $F(x,y)=F(x,y+0)$;

(3) $0\leqslant F(x,y)\leqslant 1$ for any x,y and

$$F(-\infty,x)=\lim_{x\to-\infty}F(x,y)=0;\quad F(x,-\infty)=\lim_{y\to-\infty}F(x,y)=0;$$

$$F(-\infty,-\infty)=\lim_{\substack{x\to-\infty\\y\to-\infty}}F(x,y)=0;\quad F(+\infty,+\infty)=\lim_{\substack{x\to+\infty\\y\to+\infty}}F(x,y)=1.$$

Definition 3.11 Joint probability mass function (Joint PMF)[1]

Let (X, Y) be a two-dimensional random variable, we define the joint probability mass function to be

$$p_{X,Y}(x,y)=P\{\{X=x\}\cap\{Y=y\}\}=P\{X=x,Y=y\} \tag{3.12}$$

In this setting, $p_Y(x)$ and $p_Y(y)$ are referred to as the marginal probability mass functions.

The probability distribution of two-dimensional random variable (X, Y) can also be expressed in tabular form:

$p(x,y)$	x_1	x_2	$\cdots$	x_i	$\cdots$
y_1	p_{11}	p_{21}	$\cdots$	p_{i1}	$\cdots$
y_2	p_{12}	p_{22}	$\cdots$	p_{i2}	$\cdots$
$\vdots$	$\vdots$	$\vdots$		$\vdots$	
y_j	p_{1j}	p_{2j}		p_{ij}	$\cdots$
$\vdots$	$\vdots$	$\vdots$		$\vdots$	

Similar to the distribution law of one-dimensional discrete random variables, the joint distribution law of two-dimensional discrete random variables has the following properties:

(1) $p_{ij}\geqslant 0, i,j=1,2,\cdots$;

[1] Joint probability mass function (Joint PMF): 联合概率质量函数

(2) $\sum_{i=1}^{\infty}\sum_{j=1}^{\infty}p_{ij}=1$.

3.4.2 Marginal Distribution

The joint PMF determines the probability of any event that can be specified in terms of the random variables X and Y .

Proposition 3.5 The probability $P\{(X,Y)\in A\}$ that the random pair (X,Y) lies in the set A is obtained by summing the joint PMF over pairs in A:

If A is the set of all pairs (x,y) that have a certain property, then

$$P\{(X,Y)\in A\}=\sum_{(x,y)\in A}p_{X,Y}(x,y).$$

In particular, if $A=\{(a,b)\mid a\leqslant x,b\leqslant y\}$, then

$$P\{(X,Y)\in A\}=\sum_{a\leqslant x,b\leqslant y}p_{X,Y}(a,b)=P\{X\leqslant x,Y\leqslant y\}=F(x,y),$$

which means that we can **get joint CDF from joint PMF.**

If $A_1=\{(a,b)\mid a=x,b\in R_Y\},A_2=\{(a,b)\mid a\in R_X,b=y\}$, then

$$P\{(X,Y)\in A_1\}=\sum_{a=x,b\in R_Y}p_{X,Y}(a,b)=P\{X=x\}=p_X(x),$$

$$P\{(X,Y)\in A_2\}=\sum_{a\in R_X,b=y}p_{X,Y}(a,b)=P\{Y=y\}=p_Y(y),$$

where R_X,R_Y are the sample spaces of X and Y respectively, $p_X(x),p_Y(y)$ are called **marginal PMF**[1] of X and Y respectively, which means that we can **get marginal PMF from joint PMF.**

【Example 3.13】 If the joint PMF of two-dimensional random variable (X,Y) is

Y \ X	1	2
1	0.1	0.4
2	0.3	0.2

Obtain the joint CDF of (X,Y) and the marginal PMF of X and Y.

Solution: (1) From Figure 3.5, we find that joint CDF is a piecewise function segmented according to the possible value of (X,Y), the details are as follows:

$$F_{X,Y}(x,y)=\begin{cases}0, & x<1 \text{ or } y<1,\\ 0.1, & 1\leqslant x<2,1\leqslant y<2,\\ 0.1+0.4=0.5, & x\geqslant 2,1\leqslant y<2,\\ 0.1+0.3=0.4, & 1\leqslant x<2,y\geqslant 2,\\ 0.1+0.3+0.4+0.2=1, & x\geqslant 2,y\geqslant 2.\end{cases}$$

[1] marginal PMF (marginal probability mass function): 边际 PMF

Note:

- With the exception of a slightly more complex calculation, the property of the multi—dimensional discrete random is similar to one-dimensional one, i. e, the joint CDF and joint PMF of multi-dimensional random variables can also be mutually deduced.

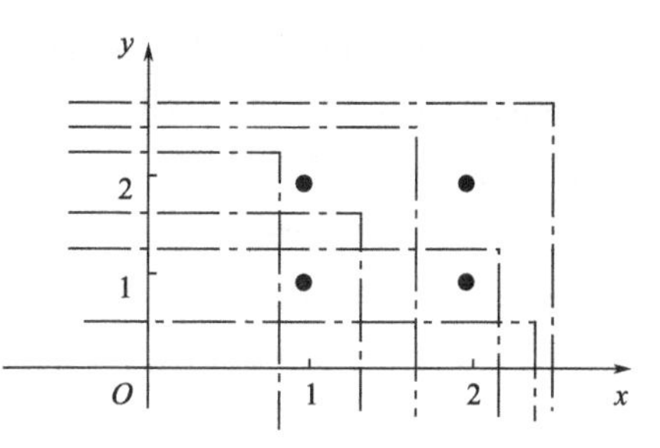

Figure 3. 5 Probability distribution map of (X, Y)

As for this example, we can find out all the possible values of (X, Y) and the corresponding probability from the segmented point of the joint CDF, so the joint PMF is as follows:

$$p_{X,Y}(1,1)=F_{X,Y}(1,1)=0.1,$$
$$p_{X,Y}(2,1)=F_{X,Y}(2,1)-F_{X,Y}(1,1)=0.5-0.1=0.4,$$
$$p_{X,Y}(1,2)=F_{X,Y}(1,2)-F_{X,Y}(1,1)=0.4-0.1=0.3,$$
$$p_{X,Y}(2,2)=F_{X,Y}(2,2)-F_{X,Y}(1,2)-F_{X,Y}(2,1)+F_{X,Y}(1,1)$$
$$=1-0.4-0.5+0.1=0.2.$$

(2) Based on definition of marginal PMF, we obtain the marginal PMF of X is

$$p_X(1)=p_{X,Y}(1,1)+p_{X,Y}(1,2)=0.4,$$
$$p_X(2)=p_{X,Y}(2,1)+p_{X,Y}(2,2)=0.6,$$

and the marginal PMF of Y is

$$p_Y(1)=p_{X,Y}(1,1)+p_{X,Y}(2,1)=0.5,$$
$$p_Y(2)=p_{X,Y}(1,2)+p_{X,Y}(2,2)=0.5.$$

Intuitively, we can get the marginal probability from the summation of the probabilities of the same row and the same column.

Y \ X	1		2	$p_Y(y)$
1	0. 1	+	0. 4	0. 5
	+		+	
2	0. 3	+	0. 2	0. 5
$p_X(x)$	0. 4		0. 6	

3. 4. 3 Conditional Distribution

In many situations, information about the observed value of one of the two variables X and Y gives information about the value of the other variable. In Example 3. 13, the probabilities of $Y=1$ and $Y=2$ are equal, but if we learn that

$X=2$, will they be still equal? The results are obviously uncertain! In fact, knowledge about X will affect the probabilities of $Y=1$, it's essentially a matter of conditional probability.

As discussed in Chapter 2, conditional probabilities are like ordinary probabilities (satisfy the three axioms) except that they refer to a new universe in which event A is known to have occurred. So, we can talk about conditional PMFs which provide the probabilities of the possible values of a random variable, conditioned on the occurrence of some event. There is nothing much new, except some new notation.

Definition 3.12 Conditional PMF[1]

Suppose we have two discrete random variables X and Y, for a fixed y, if $p_Y(y)>0$, we define

$$p_{X|Y}(x|y)=\frac{P\{X=x,Y=y\}}{P\{Y=y\}}=\frac{p_{X,Y}(x,y)}{p_Y(y)} \tag{3.13}$$

to be the conditional PMF of X under condition of $Y=y$.

Similarly, the PMF of Y under condition $X=x$ is defined to be

$$p_{Y|X}(y|x)=\frac{P\{X=x,Y=y\}}{P\{X=x\}}=\frac{p_{X,Y}(x,y)}{p_X(x)} \tag{3.14}$$

where $p_X(x)>0$.

These functions are valid PMFs for X and Y, because they satisfy two properties of PMF. Take X for example, it assigns nonnegative values to each possible x, and these values add to 1:

(1) $p_{X|Y}(x|y)=\dfrac{p_{X,Y}(x,y)}{p_Y(y)}>0$;

(2) $\sum\limits_x p_{X|Y}(x\mid y)=\sum\limits_x \dfrac{p_{X,Y}(x,y)}{p_Y(y)}=\dfrac{\sum\limits_x p_{X,Y}(x,y)}{p_Y(y)}=\dfrac{p_Y(y)}{p_Y(y)}=1.$

Note:

- From the definition of conditional PMF, it's convenient to calculate the joint PMF by marginal and conditional PMF:

$$p_{X,Y}(x,y)=p_{X|Y}(x|y)p_Y(y)=p_{Y|X}(y|x)p_X(x).$$

- It's also convenient to calculate one conditional PMF from the other conditional PMF and marginal PMF:

$$p_{X|Y}(x|y)=\frac{p_{X,Y}(x,y)}{p_Y(y)}=\frac{p_{Y|X}(y|x)p_X(x)}{p_Y(y)}.$$

---

[1] conditional PMF (conditional probability mass function): 条件概率质量函数

This method is entirely similar to the use of the multiplication rule from Chapter 2. The following examples provide an illustration.

【Example 3. 14】 Put three identical balls into three boxes with equal probability sequentially. Write X, Y as the number of balls falling into the first box and second box respectively. How to get the joint PMF of (X,Y) ? What's the probability of $X=1$ under condition of $Y=1$?

Solution: Obviously, the possible values of X, Y are $0,1,2,3$. The marginal distribution of X and the conditional distribution of Y on X are both Binominal distribution, $X\sim \mathrm{Bin}(3,1/3)$ and $Y|X=x\sim \mathrm{Bin}(3-x,1/2)$, that is

$$p_X(x)=C_3^x\left(\frac{1}{3}\right)^x\left(\frac{2}{3}\right)^{3-x},\quad p_{Y|X}(y|x)=\begin{cases}C_{3-x}^y\left(\frac{1}{2}\right)^y\left(\frac{1}{2}\right)^{3-x-y}, & 0\leqslant x+y\leqslant 3,\\ 0, & x+y>3.\end{cases}$$

Thus, the joint PMF is

$$\begin{aligned}p_{X,Y}(x,y)&=p_X(x)p_{Y|X}(y|x)\\&=\begin{cases}C_3^x\left(\frac{1}{3}\right)^x\left(\frac{2}{3}\right)^{3-x}C_{3-x}^y\left(\frac{1}{2}\right)^y\left(\frac{1}{2}\right)^{3-x-y}, & 0\leqslant x+y\leqslant 3,\\ 0, & x+y>3\end{cases}\\&=\begin{cases}\frac{1}{27}\frac{3!}{x!\ y!(3-x-y)!}, & 0\leqslant x+y\leqslant 3,\\ 0, & x+y>3.\end{cases}\end{aligned}$$

The table form of joint PMF is as follows:

Y \ X	0	1	2	3	$p_Y(y)$
0	1/27	1/9	1/9	1/27	8/27
1	1/9	2/9	1/9	0	4/9
2	1/9	1/9	0	0	2/9
3	1/27	0	0	0	1/27
$p_X(x)$	8/27	4/9	2/9	1/27	

To find the PMF of Y, we use the total probability formula

$$p_Y(y)=\sum_x p_{X,Y}(x,y)$$

The results are shown in the table above.

The probability of $X=1$ under condition of $Y=1$ can be obtained by

$$p_{X|Y}(y|x)=\frac{p_{X,Y}(x,y)}{p_Y(y)}$$

i. e.

$$p_{X|Y}(1|1)=\frac{p_{X,Y}(1,1)}{p_Y(1)}=\frac{2/9}{4/9}=\frac{1}{2}.$$

3.4.4 Independence of Discrete Random Variables

We now discuss concepts of independence related to random variables. These concepts are analogous to the concepts of independence between events discussed in Chapter 2. The independence of two random variables is similar to the independence of two events.

When we say event A and event B are independent of each other, there must have

$$P(AB)=P(A)P(B).$$

And because

$$P(AB)=P(A|B)P(B)=P(B|A)P(A),$$

we get that

$$P(A)=P(A|B) \text{ and } P(B)=P(B|A),$$

which means that knowing the occurrence of the conditioning event tells us nothing about the target event.

Now suppose that event A is "$X=x$", event B is "$Y=y$" for any given x, y, if

$$P\{X=x,Y=y\}=P\{X=x\}P\{Y=y\}$$

we say that the random variable X is independent of random variable Y.

Definition 3.13 The two random variables X and Y are **independent** if

$$p_{X,Y}(x,y)=p_X(x)p_Y(y), \text{ for all } x,y. \tag{3.15}$$

Finally, the formula $p_{X,Y}(x,y)=p_{X|Y}(x|y)p_Y(y)$ shows that "independence" is equivalent to "all conditional probabilities are equal to unconditional probabilities",

$$p_{X|Y}(x|y)=p_X(x) \text{ for all } y \text{ with } p_Y(y)>0 \text{ and all } x.$$

Intuitively, independence means that the experimental value of Y tells us nothing about the value of X, and vice versa.

【Example 3.15】 Let the joint PMF of X and Y be shown in the following table,

X \ Y	-1	0	2
1/2	2/20	1/20	2/20
1	2/20	1/20	2/20
2	4/20	2/20	4/20

Prove that X and Y are independent.

Proof: Firstly, obtain the marginal PMF of X and Y according to the joint PMF

$$\begin{cases} p_X(1/2)=2/20+1/20+2/20=1/4 \\ p_X(1)=2/20+1/20+2/20=1/4 \\ p_X(2)=4/20+2/20+4/20=1/2 \end{cases} \quad \text{and} \quad \begin{cases} p_Y(-1)=2/20+2/20+4/20=2/5 \\ p_Y(0)=1/20+1/20+2/20=1/5 \\ p_Y(2)=2/20+2/20+4/20=2/5 \end{cases}$$

By verifying all the equations $p_{X,Y}(x,y)=p_X(x)p_Y(y)$ for all x,y, we draw that X and Y are independent.

3.4.5 Derived Distributions of Multiple Discrete Random Variables

When there are multiple random variables of interest, it is possible to generate new random variables by considering functions involving several of these random variables. In particular, a function $Z=g(X,Y)$ of the random variables X and Y defines another random variable. Its PMF can be calculated from the joint PMF $p_{X,Y}(x,y)$ according to

$$p_Z(z)=\sum_{\{(x,y)\mid g(x,y)=z\}} p_{X,Y}(x,y). \tag{3.16}$$

In particular, when X and Y are independent,

$$p_Z(z)=\sum_{\{(x,y)\mid g(x,y)=z\}} p_X(x)p_Y(y). \tag{3.17}$$

【Example 3.16】 It's known that two random variables X and Y are independent, $X\sim\text{Poisson}(\lambda_1)$, $Y\sim\text{Poisson}(\lambda_2)$. To get the PMF of $Z=X+Y$.

Solution: Because both X and Y follows Poisson distribution, the possible values of them are any nonnegative integers, so the possible values of $Z=X+Y$ are also nonnegative integers.

$$p_Z(z)=\sum_{\{(x,y)\mid x+y=z\}} p_{X,Y}(x,y)=\sum_{\{(x,y)\mid x+y=z\}} p_X(x)p_Y(y)=\sum_{i=0}^{z} p_X(i)p_Y(z-i)$$

$$=\sum_{i=0}^{z}\frac{\lambda_1{}^i}{i!}e^{-\lambda_1}\frac{\lambda_2{}^{z-i}}{(z-i)!}e^{-\lambda_2}=\frac{e^{-(\lambda_1+\lambda_2)}}{z!}\sum_{i=0}^{z}C_z^i\lambda_1{}^i\lambda_2{}^{z-i}=\frac{(\lambda_1+\lambda_2)^z}{z!}e^{-(\lambda_1+\lambda_2)}.$$

That is, $Z=X+Y\sim\text{Possion}(\lambda_1+\lambda_2)$. This conclusion is also called **additivity of Poisson distribution.**

Exercises

1. There are two kinds of famous wine with very similar taste and color, four glasses each. If someone picks out four of them and they are all the same wine, which is considered as a successful experiment.

   (1) Someone randomly guesses, what is the probability that he will succeed in the experiment at one time?

(2) Someone claims that he can distinguish between two kinds of wine by tasting. He experimented 10 times in a row and succeeded three times. Try to infer whether he is right or whether he does have the ability to distinguish (assuming that the experiments are independent of each other).

2. A mail-order computer business has six telephone lines. Let X denote the number of lines in use at a specified time. Suppose the PMF of X is as given in the accompanying table.

X	0	1	2	3	4	5	6
$p_X(x)$	0.1	0.15	0.2	0.25	0.2	0.06	0.04

Calculate the probability of each of the following events.

(1) {at most three lines are in use};

(2) {fewer than three lines are in use};

(3) {at least three lines are in use};

(4) {between two and five lines, inclusive, are in use};

(5) {between two and four lines, inclusive, are not in use};

(6) {at least four lines are not in use}.

3. Show that the CDF $F(x)$ is a non-decreasing function; that is, $x_1 < x_2$ implies that $F(x_1) < F(x_2)$. Under what condition will $F(x_1) = F(x_2)$?

4. Suppose, for a lottery scratch card, probabilities of different prizes are as follows:

Prize	£0	£2	£10	£100
Probability	0.689	0.300	0.010	0.001

If a scratch card costs £1, and you buy one card, tabulate the PMF and CDF of your profit.

5. A branch of a certain bank in New York City has six ATMs. Let X represent the number of machines in use at a particular time of day. The CDF of X is as follows:

$$F_X(x)=\begin{cases} 0, & x<0, \\ 0.06, & 0 \leqslant x<1, \\ 0.19, & 1 \leqslant x<2, \\ 0.39, & 2 \leqslant x<3, \\ 0.67, & 3 \leqslant x<4, \\ 0.92, & 4 \leqslant x<5, \\ 0.97, & 5 \leqslant x<6, \\ 1, & 6 \leqslant x. \end{cases}$$

Calculate the following probabilities directly from the CDF $F_X(x)$:

(1) $P\{X=2\}$; (2) $P\{X>3\}$; (3) $P\{2\leqslant X\leqslant 5\}$; (4) $P\{2<X<5\}$.

6. An insurance company offers its policyholders a number of different premium payment options. For a randomly selected policyholder, let $X=$ the number of months between successive payments. The CDF of X is as follows:

$$F_X(x)=\begin{cases}0, & x<1,\\ 0.30, & 1\leqslant x<3,\\ 0.40 & 3\leqslant x<4,\\ 0.45 & 4\leqslant x<6,\\ 0.60, & 6\leqslant x<12,\\ 1, & 12\leqslant x.\end{cases}$$

(1) What is the PMF of X?

(2) Using just the CDF, compute $P\{3\leqslant x\leqslant 6\}$ and $P\{X\geqslant 4\}$.

7. A TV chef claims his free-range chickens taste better than battery-farmed chickens. 10 people are given a sample of each to taste, without knowing which is which, and are asked to state which sample they prefer. It is suggested that the participants cannot actually taste the difference, and are effectively choosing which sample they prefer at random. If this is true,

(1) What probability distribution would you use to describe the number of people who say they prefer the free-range chicken?

(2) Using your distribution in (1), calculate the probability that the number of people who say they prefer the free-range chicken is: (a) not 5; (b) no more than 8.

8. In a production line, it is estimated that 1 in every 200 items will be faulty. At the quality control stage, each item is visually inspected, and it is believed that there is a 90% chance of detecting a faulty item, and a 5% chance of mistakenly declaring a non-faulty item as being faulty. In a batch of 10 items, what is the probability of two items being declared faulty at the quality control stage?

9. The mode of a discrete random variable X with PMF $p(x)$ is that value x^* for which $p(x)$ is largest (the most probable x value). Let $X\sim \mathrm{Bin}(n,p)$. By considering the ratio $p(x+1;n,p)/p(x;n,p)$, show that $p(x;n,p)$ increase with x as long as $x<np-(1-p)$. Conclude that the mode x^* is the integer satisfying $(n+1)p-1<x^*<(n+1)p$.

10. An article reports that 1 in 200 people carry the defective gene that causes inherited colon cancer. In a sample of 1000 individuals, what is the approximate distribution of the number who carry this gene? Use this distribution to calculate the approximate probability that

(1) Between 5 and 8 (inclusive) carry the gene;

(2) At least 8 carry the gene

11. Of the people passing through an airport metal detector, 0.5% activate it; let X = the number among a randomly selected group of 500 who activate the detector.

(1) What is the (approximate) PMF of X ?

(2) Compute $P\{X=5\}$;

(3) Compute $P\{X \geqslant 5\}$.

12. Let X be the number of material anomalies occurring in a particular region of an aircraft gas-turbine disk. An article proposes a Poisson distribution for X with parameter 4.

(1) Compute both $P\{X \leqslant 4\}$ and $P\{X < 4\}$;

(2) Compute $P\{4 \leqslant X \leqslant 8\}$;

(3) Compute $P\{X \geqslant 8\}$.

13. Let joint PMF of X, Y is as follows:

Y \ X	1	2	3
1	1/6	1/9	1/18
2	1/3	α	β

What values do α and β take to make X and Y independent of each other?

14. A service station has both self-service and full-service islands. On each island, there is a single regular unleaded pump with two hoses. Let X denote the number of hoses being used on the self-service island at a particular time, and let Y denote the number of hoses on the full-service island in use at that time. The joint PMF of X and Y appears in the accompanying tabulation.

Y \ X	0	1	2
0	0.10	0.04	0.02
1	0.08	0.20	0.06
2	0.06	0.14	0.30

(1) What is $P\{X=1, Y=1\}$?

(2) Compute $P\{X \leqslant 1, Y \leqslant 1\}$.

(3) Give a word description of the event $\{X \neq 0, Y \neq 0\}$, and compute the probability of this event.

(4) Compute the marginal PMF of X and of Y. Using $p_X(x)$, what is $P\{X \leqslant 1\}$?

(5) Are X and Y independent random variables? Explain.

15. Let X denote the number of Canon SLR cameras sold during a particular week by a certain store. The PMF of X is

X	0	1	2	3	4
$p_X(x)$	0.1	0.2	0.3	0.25	0.15

Sixty percent of all customers who purchase these cameras also buy an extended warranty. Let Y denote the number of purchasers during this week who buy an extended warranty.

(1) What is $P\{X=4, Y=2\}$? [Hint: This probability equals $P\{Y=2 \mid X=4\} \cdot P\{X=4\}$; now think of the four purchase as four trials of a binomial experiment, with success on a trial corresponding to buying an extended warranty.]

(2) Calculate $P\{X=Y\}$.

(3) Determine the joint PMF of X and Y and then the marginal PMF of Y.

16. Two fair six-sided dice are tossed independently. Let M=the maximum of the two tosses [so $M(1,5)=5$, $M(3,3)=3$, etc.].

(1) What is the PMF of M? [Hint: First determine $p(1)$, then $p(2)$, and so on.]

(2) Determine the CDF of M.

17. Let the PMFs of X and Y are as follows:

X	-1	0	1
$p_X(x)$	0.25	0.5	0.25

Y	0	1
$p_Y(y)$	0.5	0.5

It's known that $P\{XY=0\}=1$, Try to find the PMF of random variable $Z=\max\{X, Y\}$.

Chapter 4 Continuous Random Variables

In Chapter 3 we concentrate on the development of probability distributions for discrete random variables. In this chapter, we continue with the study of random variables, but now consider the case with a continuous range of possible experimental values. Examples of continuous random variables are

- the temperature (in degree Celsius) in Nanjing tomorrow;
- your current blood pressure level (in millimetres of mercury);
- the amount of time a randomly selected customer spends waiting for a haircut before his/her haircut commences.

In practice, we can limit our measuring instruments restrict to a discrete world. but it's usually more convenient to treat them as continuous. Partly because continuous models often approximate real-world situations very well, and partly because continuous mathematics (the calculus) is frequently easier to work with than mathematics of discrete variables and distributions.

All of the concepts and methods introduced in Chapter 3, such as probability distribution, joint probability distributions, independence, derived distributions have continuous counterparts. Developing and interpreting these counterparts is the subject of this chapter.

4.1 Continuous Random Variable

4.1.1 Continuous Probability Distribution

Suppose the variable X of interest is the depth of a lake at a randomly chosen point on the surface. Let M equals the maximum depth (in meters), so that any number in the interval $[0,M]$ is a possible value of X.

If we write $P\{X=x\}=k$ for some constant value k, what value would k be? In addition, there are uncountably many different x in the interval $[0,M]$, so how can we sum k an infinite number of times and get 1? Clearly, this isn't going to work.

But if we "discretize" X by taking the integer value of the true depth, then possible values are nonnegative integers less than or equal to M. The resulting discrete distribution of depth can be pictured using a probability histogram. If we draw the histogram so that the area of the rectangle above any possible integer k is the proportion of the lake whose depth is (to the nearest meter) k, then the total area of all rectangles is 1. A possible histogram appears in Figure 4. 1(a). If depth is measured much more accurately and the same measurement axis as in Figure 4. 1 (a) is used, each rectangle in the resulting probability histogram is much narrower [see Figure 4. 1(b)], the histogram has a much smoother appearance than in Figure 4. 1(a), though the total area of all rectangles is still 1. If we continue in this way to measure depth more and more finely, as shown in the histogram of Figure 4. 1(c). the resulting sequence of histograms approaches a smooth curve [see Figure 4. 1(c)]. Because for each histogram the total area of all rectangles equals 1, the total area under the smooth curve is also 1. The probability that the depth at a randomly chosen point is between a and b is just the area under the smooth curve between a and b. It is exactly a smooth curve of the type pictured in Figure 4. 1 (c) that specifies a continuous probability distribution.

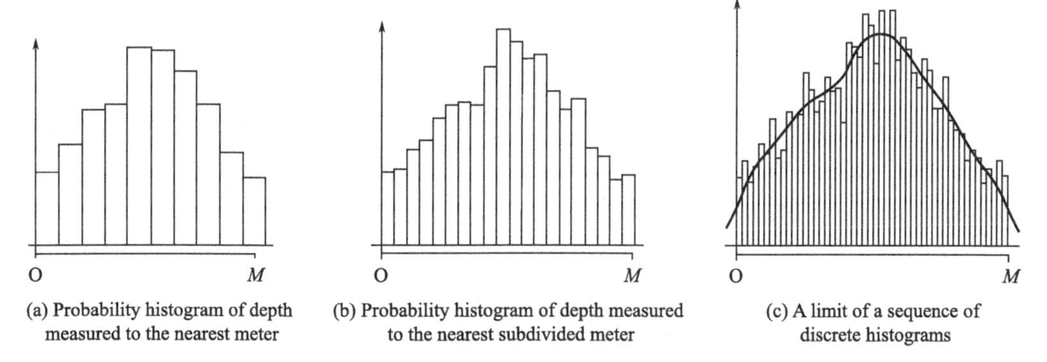

Figure 4. 1 Probability histogram of depth

Probability Density Function

Definition 4. 1 Let X be a continuous r. v. Then the **probability density function (PDF)**[1] of X is a function such that for any two numbers a and b with $a \leqslant b$

$$P\{a \leqslant X \leqslant b\} = \int_a^b f(x)\mathrm{d}x. \tag{4.1}$$

[1] probability density function (PDF)：概率密度函数

The graph of $f(x)$ is often referred to as the **density curve.**

That is, the probability that X takes on a value in the interval $[a,b]$ is the area above this interval and under the density curve. (see Figure 4.2)

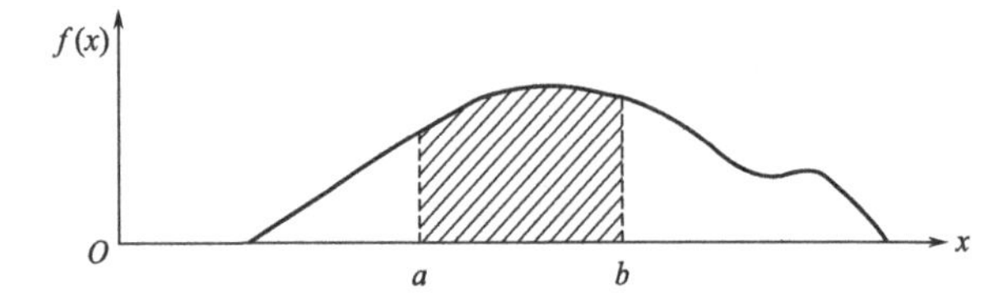

Figure 4.2 Probability of X between a and b is the area under the density curve between a and b

Proposition 4.1 As a probability density function, the function $f_X(x)$ must be nonnegative, and must also satisfy the normalization equation

(1) $f_X(x) \geqslant 0, x \in R$;

(2) $\int_{-\infty}^{+\infty} f_X(x)\mathrm{d}x = P\{-\infty < X < +\infty\} = 1$.

Note:

• It is worth to note that for event of the continuous random variable equals some value, the probability of which is the area under a density curve that lies above any single value is zero, i. e. :

$$P\{X=c\} = \int_c^c f(x)\mathrm{d}x = \lim_{\varepsilon \to 0}\int_{c-\varepsilon}^{c+\varepsilon} f(x)\mathrm{d}x = 0.$$

From above, we can see that it is different between discrete and continuous random variables. When X is a discrete random variable, each possible value is assigned positive probability, but it is not true for continuous random variable. Therefore, when X is continuous random variable, the following probabilities shown in equation (4.2) are equivalent, that is, the probability that X lies in some interval between a and b does not depend on whether the lower limit a or the upper limit b is included in the probability calculation:

$$P\{a \leqslant X \leqslant b\} = P\{a < X < b\} = P\{a < X \leqslant b\} = P\{a \leqslant X < b\}. \tag{4.2}$$

Cumulative Distribution Functions

We know that the cumulative distribution function $F(x)$ is the probability that the random variable is less than or equal to x, but instead of a summation of probability for a discrete random variable, it is calculated by integrating the PDF $f(x)$ between $-\infty$ and x for a continuous random variable.

Definition 4.2 **Cumulative Distribution Function (CDF)**[1] $F(x)$ for a continuous r. v X is defined for every number x by

$$F(x)=P\{X\leqslant x\}=\int_{-\infty}^{x}f(t)\mathrm{d}t. \tag{4.3}$$

For each x, $F(x)$ is the area under the density curve to the left of x, obviously, **$F(x)$ increases smoothly as x increases.** (see Figure 4.3)

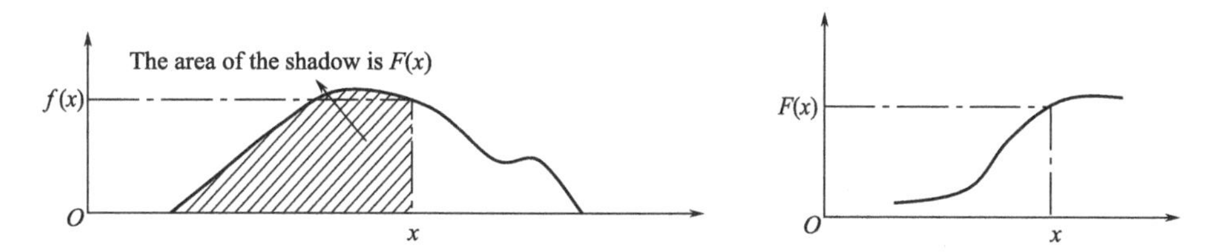

Figure 4.3 PDF of continuous Random variables and the corresponding CDF

Mutual Deduction between $f(x)$ and $F(x)$

For discrete random variable X, the PDF is obtained from the CDF by taking the difference between two $F(x)$ values. For continuous random variable, we use derivative to get PDF.

Proposition 4.2 If X is a continuous r. v with PDF $f(x)$ and CDF $F(x)$, then at every x at which the derivative $F'(x)$ exists, $F'(x)=f(x)$.

The cumulative distribution function $F(x)$ and the probability density function $f(x)$ can be deduced mutually, and each of them can represent the distribution characteristics of continuous random variables.

【Example 4.1】 (From PDF to CDF)

Let x be a random variable with the probability density function $f(x)$ as follows:

$$f(x)=\begin{cases} kx, & 0\leqslant x<3, \\ 2-\dfrac{x}{2}, & 3\leqslant x<4, \\ 0, & \text{otherwise}. \end{cases}$$

Find the cumulative distribution function of X and the probability of X in the interval $[1,3.5]$.

Solution: From the fact that $\int_{-\infty}^{+\infty}f_X(x)\mathrm{d}x=1$, we get

$$\int_0^3 kx\,\mathrm{d}x+\int_3^4\left(2-\frac{x}{2}\right)\mathrm{d}x=1,$$

---

[1] Cumulative Distribution Function (CDF)：累积分布函数

so $k=\frac{1}{6}$.

From $F(x)=\int_{-\infty}^{x} f(t)\mathrm{d}t$, we get

$$F(x)=\begin{cases}0, & x<0,\\ \int_0^x \frac{x}{6}\mathrm{d}x, & 0\leqslant x<3,\\ \int_0^3 \frac{x}{6}\mathrm{d}x+\int_3^x\left(2-\frac{x}{2}\right)\mathrm{d}x, & 3\leqslant x<4,\\ 1, & x\geqslant 4\end{cases}=\begin{cases}0, & x<0,\\ \frac{x^2}{12}, & 0\leqslant x<3,\\ -\frac{x^2}{4}+2x-3, & 3\leqslant x<4,\\ 1, & x\geqslant 4.\end{cases}$$

So, $P\{1<x\leqslant 3.5\}=F(3.5)-F(1)=0.854$.

【Example 4.2】 (From CDF to PDF)

Let $F(x)$ is the CDF of a continuous random variable X,

$$F(x)=\begin{cases}A\mathrm{e}^x, & x<0,\\ B, & 0\leqslant x<1,\\ 1-A\mathrm{e}^{-(x-1)}, & x\geqslant 1.\end{cases}$$

Find the PDF of X, and the probability of X in the interval $(1/3,+\infty)$.

Solution: Since **$F(x)$ is continuous function**, it must be continuous at points $x=0$ and $x=1$, that is,

$$\begin{cases}\lim\limits_{x\to 0^-}F(x)=\lim\limits_{x\to 0^-}A\mathrm{e}^x=A=F(0)=B;\\ \lim\limits_{x\to 1^-}F(x)=\lim\limits_{x\to 1^-}B=F(1)=1-A.\end{cases}\tag{4.4}$$

From equation (4.4), we get $A=B=1/2$.

So, the PDF is as follows,

$$f(x)=F'(x)=\begin{cases}A\mathrm{e}^x, & x<0,\\ 0, & 0\leqslant x<1,\\ A\mathrm{e}^{-(x-1)}, & x\geqslant 1\end{cases}=\begin{cases}\frac{1}{2}\mathrm{e}^x, & x<0,\\ 0, & 0\leqslant x<1,\\ \frac{1}{2}\mathrm{e}^{-(x-1)}, & x\geqslant 1.\end{cases}$$

Then we can get the probability based on two functions respectively:

$$P\{X>\frac{1}{3}\}=1-P\{X\leqslant\frac{1}{3}\}=1-F\left(\frac{1}{3}\right)=1-\frac{1}{2}=\frac{1}{2};$$

or

$$P\{X>\frac{1}{3}\}=\int_{\frac{1}{3}}^{+\infty}f(x)\mathrm{d}x=\int_{\frac{1}{3}}^{0}0\mathrm{d}x+\int_1^{+\infty}\frac{\mathrm{e}^{-(x-1)}}{2}\mathrm{d}x=\frac{1}{2}.$$

4.1.2 Some Important Continuous Distribution

The uniform distribution

The uniform distribution is used to describe a random variable that is constrained to lie in some interval $[a,b]$, and all subintervals of the same length are equal likely. Its PDF has the form

$$f(x)=\begin{cases} c, a\leqslant x\leqslant b, \\ 0, \text{otherwise.} \end{cases}$$

where c is a constant. For $f(x)$ to satisfy the normalization property, we must have

$$1=\int_{-\infty}^{+\infty} f(x)\mathrm{d}x=\int_{a}^{b} c\,\mathrm{d}x=c(b-a).$$

So that $c=\dfrac{1}{b-a}$.

Definition 4.3 **(Uniform distribution)**[1] Random variable X is considered to follow **uniform distribution** over the interval $[a, b]$, denoted by $X\sim U[a,b]$, if its PDF is

$$f(x)=\begin{cases} 1/(b-a), & a\leqslant x\leqslant b, \\ 0, & \text{otherwise.} \end{cases} \tag{4.5}$$

Figure 4.4 illustrates the PDF curve of a uniform distribution.

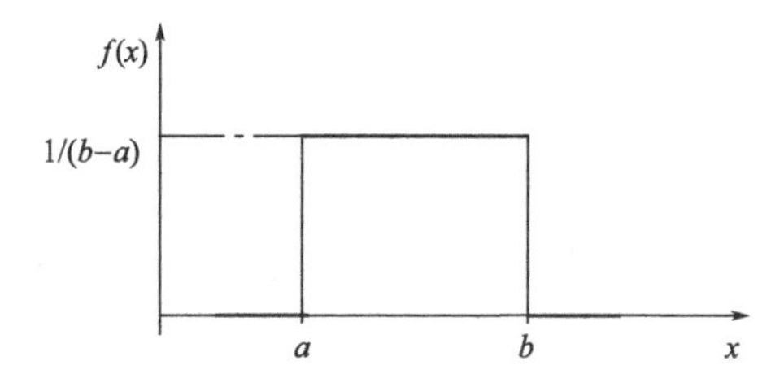

Figure 4.4 The PDF of a uniform random variable

Since $F(x)$ is the integral of $f(x)$ from minus infinity to x, we have

$$F(x)=\int_{-\infty}^{x} f(t)\mathrm{d}t=\begin{cases} \int_{-\infty}^{x} 0\mathrm{d}t=0, & x<a, \\ \int_{-\infty}^{a} 0\mathrm{d}t+\int_{a}^{x} \dfrac{1}{b-a}\mathrm{d}t=\dfrac{x-a}{b-a}, & a\leqslant x<b, \\ \int_{-\infty}^{a} 0\mathrm{d}t+\int_{a}^{b} \dfrac{1}{b-a}\mathrm{d}t+\int_{b}^{x} 0\mathrm{d}t=1, & x\geqslant b. \end{cases}$$

Proposition 4.3 If random variable X follows uniform distribution $X\sim U[a,b]$, then it's CDF is

[1] Uniform distribution：均匀分布

$$F_X(x)=\begin{cases}0, & x<a,\\ \dfrac{x-a}{b-a}, & a\leqslant x<b,\\ 1, & x\geqslant b.\end{cases} \tag{4.6}$$

The uniform distribution is an important concept in probability theory, but it is less useful for modelling uncertainty in the real world, because it is not often plausible in real situations that all intervals of the same width are equally likely.

The exponential distribution

【Example 4.3】 Suppose the number of earthquakes N_t in an interval $[0,t]$ follows Poisson distribution $N_t\sim\text{Poisson}(\lambda t)$ (for any value of t). Let T be the time until the first earthquake. Which probability distribution does random variable T follow? (Hint: for $T>t$ to occur, what value must N_t take?)

Solution: In fact, event "the time to wait for the next occurrence Y is longer than T, i.e $Y>T$" is equivalent to event "the number of events within $[0,T]$ is zero", that is, $P\{Y>T\}=P\{N_t=0\}$ Since $N_t\sim\text{Poisson}(\lambda t)$, so

$$P\{Y>T\}=P\{N_t=0\}=\frac{\lambda T^0 \mathrm{e}^{-\lambda T}}{0!}=\mathrm{e}^{-\lambda T}.$$

Therefore

$$P\{Y\leqslant T\}=1-\mathrm{e}^{-\lambda T}=F_Y(T).$$

Taking the derivative to both sides of equation, we get

$$f_Y(T)=\lambda \mathrm{e}^{-\lambda T}.$$

Here, we call the waiting time T follows the exponential distribution of parameter λ.

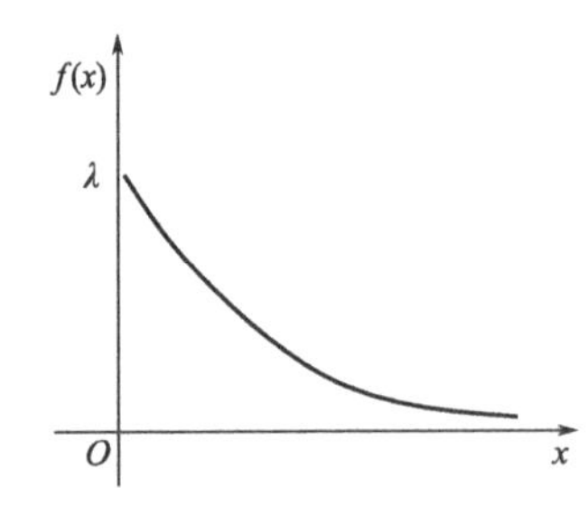

Figure 4.5 The PDF of an exponential random variable

Definition 4.4 **(Exponential distribution)**[1] Random variable X is considered to follow **Exponential distribution** with rate parameter λ ($\lambda>0$), denoted by $X\sim\text{Exp}(\lambda)$, if its PDF is

$$f(x)=\begin{cases}\lambda \mathrm{e}^{-\lambda x}, & x>0\\ 0, & \text{otherwise}\end{cases} \tag{4.7}$$

Figure 4.5 illustrates the PDF curve of a exponential distribution.

Proposition 4.4 If random variable X follows Exponential distribution $X\sim\text{Exp}(\lambda)$, then it's CDF is

$$F_X(x)=\begin{cases}1-\mathrm{e}^{-\lambda x}, & x>0,\\ 0, & \text{otherwise.}\end{cases} \tag{4.8}$$

---

[1] Exponential distribution: 指数分布

An exponential random variable can be a very good model for the amount of time until a piece of equipment breaks down, until a light bulb burns out, or until an accident occurs. For example, if a patient with heart disease is given a drug, then "the time until the patient's next heart attack" follows exponential distribution, if a new car is bought, then "the miles the car is driven before it has its first breakdown" also follows exponential distribution.

【Example 4.4】 The service life of an electronic computer X follows the exponential distribution with parameter $\lambda=0.0001$.

(1) To find the probability that the computer can be used for more than 10000 hours;

(2) If it is known that the computer has been used for 10000 hours, what is the probability that it can be used for another 10000 hours.

Solution:

Let A denotes the event "computer can last more than 10000 hours"; B denotes the event "computer can last more than 20000 hours".

Since $X\sim\text{Exp}(0.0001)$, i.e. $f(x)=\begin{cases}0.0001e^{-0.0001x}, & x>0\\ 0, & x\leqslant 0\end{cases}$, we have

$$P(A)=P\{X\geqslant 10000\}=\int_{10000}^{+\infty}0.0001e^{-0.0001x}\,dx=e^{-1}.$$

Similarly, $P(B)=P\{X\geqslant 20000\}=\int_{20000}^{+\infty}0.0001e^{-0.0001x}\,dx=e^{-2}$.

The occurrence of B must lead to the occurrence of A, that is, $B\subset A$, i.e. $AB=B$, so

$$P(B|A)=P(AB)/P(A)=P(B)/P(A)=e^{-2}/e^{-1}=e^{-1}.$$

The results show that under the condition that the computer has been used for 10000 hours, the conditional probability that the computer has been used for more than 20000 hours is equal to the unconditional probability that it can be used for more than 10000 hours, that is, the computer has no memory of the events for which it has been used for 10000 hours. This result is not accidental, it reveals an important property of exponential distribution-memoryless.

Proposition 4.5 The random variable X, which follows exponential distribution, is **memoryless**[1], that is, for any $s,t>0$, we have

$$P\{X>s+t\,|\,X>s\}=P\{X>t\}. \tag{4.9}$$

In fact, $P\{X>s+t\,|\,X>s\}=\dfrac{P\{X>s+t,X>s\}}{P\{X>s\}}=\dfrac{P\{X>s+t\}}{P\{X>s\}}$

$$=\frac{1-F(s+t)}{1-F(s)}=\frac{e^{-\lambda(s+t)}}{e^{-\lambda s}}=e^{-\lambda t}=P\{X>t\}.$$

[1] memoryless：无记忆性

The Normal Distribution

The normal distribution is the most important one in all of probability and statistics. Many numerical populations have distributions that can be fit very closely by an appropriate normal curve. For example, the heights of females in a particular age group, can be well represented with a normal distribution, 'measurement errors' are often assumed as normally distributed by scientists, changes in stock prices in finance are commonly modeled as normal distributions (though not always sensibly!). In addition, even when individual variables themselves are not normally distributed, sums and averages of the variables will under suitable conditions have approximately a normal distribution, this is the content of the Central Limit Theorem discussed in the next chapter.

Definition 4.5 Random variable X is considered to be **normal or gaussian distribution**[1] with parameters μ and σ^2, denoted by $X \sim N(\mu, \sigma^2)$, if it's PDF is

$$f_X(x) = \frac{1}{\sqrt{2\pi}\sigma} e^{-(x-\mu)^2/2\sigma^2}. \tag{4.10}$$

Figure 4.6 illustrates the PDF curves with different parameters, which shows that the changing of parameter μ moves horizontally the density curve without changing the shape of the curve, so it is referred to as a **location parameter**[2], the changing of parameter σ^2 stretches or compresses the curve horizontally without changing the basic shape, so it is referred to as a **scale parameter**[3]. A large value of σ^2 corresponds to a density curve that is quite spread with $x=\mu$ as its symmetry axis, whereas a small value of σ^2 yields a highly concentrated curve.

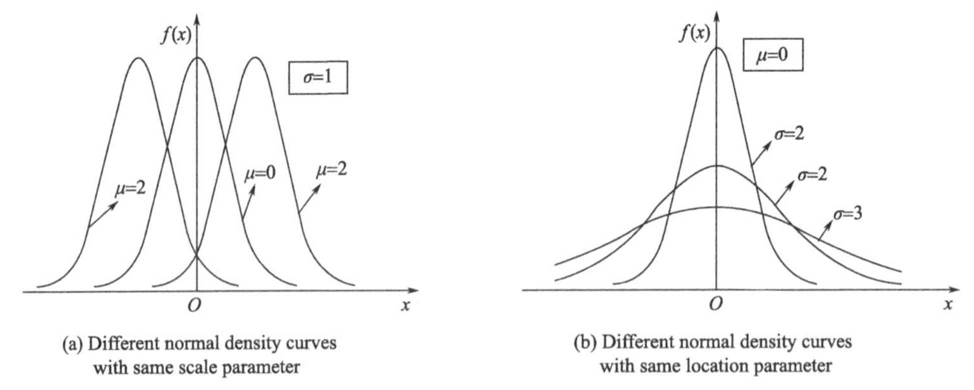

Figure 4.6 Different normal density curves

[1] normal or gaussian distribution：正态分布或高斯分布

[2] location parameter：位置参数

[3] scale parameter：尺度参数

In particular, when $\mu=0$ and $\sigma^2=1$, the random variable is called to be standard normal.

Definition 4.6 Random variable X is considered to follow **standard normal distribution**[1] with $\mu=0$ and $\sigma^2=1$, denoted by $X\sim N(0,1)$, if it's PDF is

$$f(x)=\frac{1}{\sqrt{2\pi}}e^{-x^2/2}. \tag{4.11}$$

Obviously, the PDF of normal distribution curve is a symmetric about $x=0$ and the area under the curve between $-\infty$ and $-x$ is the same as the area under the curve between x and ∞, i. e. random variables with standard normal distribution have equal probabilities of less than zero and greater than zero, and the probability of being less than $-x$ and greater than x is also equal. (see Figure 4.7) We often denote the CDF of the standard normal distribution random variables by $\Phi(x)$.

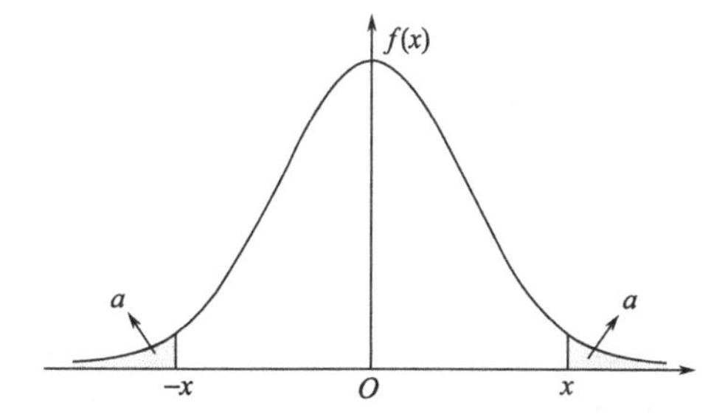

Figure 4.7 The PDF of normal distribution curve is a symmetric about $x=0$ and the area under the curve between $-\infty$ and $-x$ is the same as the area under the curve between x and ∞

Proposition 4.6 If random variable X follows **standard normal distribution** $X\sim N(0,1)$, then it's CDF $\Phi(x)$ satisfies the following properties

$$\begin{cases}\Phi(0)=0.5,\\ \Phi(-x)=1-\Phi(x).\end{cases} \tag{4.12}$$

Since $F(x)$ is the integral of $f(x)$ from minus infinity to x, we can write the CDF of X as

$$F_X(x)=\int_{-\infty}^{x}\frac{1}{\sqrt{2\pi}\sigma}e^{-(x-\mu)^2/2\sigma^2}dx.$$

But there isn't standard integration techniques can be used to accomplish this, so we have to calculate it using numerical techniques. People have calculated values of the $\Phi(x)$ for standard normal distribution when x is greater than or equal to

[1] standard normal distribution：标准正态分布

zero and tabulated them in Table 2 in appendix. Then the computation of $P\{a \leqslant X \leqslant b\} = \Phi(b) - \Phi(a)$ when X is standard normal can be got by looking up the table.

【Example 4.5】 Let's determine the following standard normal probabilities:

$P\{X \leqslant 1.25\}, P\{X > 1.25\}, P\{X \leqslant -1.25\}, P\{-0.38 \leqslant X \leqslant -1.25\}, P\{X \leqslant 5\}$.

Solution:

$P\{X \leqslant 1.25\} = \Phi(1.25) = 0.8944$, the probability value can be directly found through Table 2 in appendix.

$$P\{X > 1.25\} = 1 - P\{X \leqslant 1.25\} = 1 - \Phi(1.25) = 0.1056;$$

Because $\Phi(-x) = 1 - \Phi(x)$, so $P\{X \leqslant -1.25\} = \Phi(-1.25) = 1 - \Phi(1.25) = 0.1056$;

$P\{-0.38 \leqslant X \leqslant -1.25\} = \Phi(-1.25) - \Phi(-0.38) = 1 - \Phi(1.25) - [1 - \Phi(0.38)] = 0.5424$ (see Table 2 in appendix)

$P\{X \leqslant 5\} = \Phi(5)$, but we can not find this probability in the table because the biggest value we can find is 3.49 and $\Phi(3.49) = 0.998$ which is almost 1, because $\Phi(x)$ is increasing function, so $\Phi(3.49) < \Phi(5) \approx 1$.

The standard normal distribution is a special case of normal distributions, but in general, parameters of normal distributions can be any values. How to calculate the probability of non-standard normal distribution in a certain interval?

In fact, we can establish the relationship between the standard and non-standard normal distribution, so that we can calculate the probability by looking up Table 2 in appendix.

Theorem 4.1 Relationship between a standard normal random variable Z and a 'general' normal random variable X:

(1) Given $Z \sim N(0,1)$, we can obtain $X \sim N(\mu, \sigma^2)$ by transformation:

$$X = \mu + \sigma Z;$$

(2) Given $X \sim N(\mu, \sigma^2)$, we can **standardize** X to $Z \sim N(0,1)$ by transformation:

$$Z = (X - \mu)/\sigma.$$

Proof:

(1) Given $Z \sim N(0,1)$, we can calculate the CDF of X:

$$F_X(x) = P\{X \leqslant x\} = P\{\mu + \sigma Z \leqslant x\} = P\{Z \leqslant (x-\mu)/\sigma\} = \int_{-\infty}^{(x-\mu)/\sigma} \frac{1}{\sqrt{2\pi}} e^{-\frac{z^2}{2}} dz.$$

Finding the derivative of $F_X(x)$ to get the PDF of random variable X:

$$f_X(x) = [F_X(x)]' = \frac{1}{\sqrt{2\pi}} e^{-\frac{((x-\mu)/\sigma)^2}{2}} ((x-\mu)/\sigma)' = \frac{1}{\sqrt{2\pi}\sigma} e^{-\frac{(x-\mu)^2}{2\sigma^2}}.$$

In the form of PDF, it is clear that X follows the normal distribution with parameters μ and σ^2, i. e. $X \sim N(\mu, \sigma^2)$.

(2) Given $X \sim N(\mu, \sigma^2)$, we can calculate the CDF of Z:

$$F_Z(z)=P\{Z\leqslant z\}=P\{(X-\mu)/\sigma\leqslant z\}=P\{X\leqslant \mu+\sigma z\}=\int_{-\infty}^{\mu+\sigma z}\frac{1}{\sqrt{2\pi}\sigma}\mathrm{e}^{-\frac{(x-\mu)^2}{2\sigma^2}}\mathrm{d}x.$$

Finding the derivative of $F_Z(z)$ to get the PDF of random variable Z:

$$f_Z(z)=[F_Z(z)]'=\frac{1}{\sqrt{2\pi}\sigma}\mathrm{e}^{-\frac{(\mu+\sigma z-\mu)^2}{2\sigma^2}}(\mu+\sigma z)'=\frac{1}{\sqrt{2\pi}}\mathrm{e}^{-\frac{z^2}{2}}.$$

【Example 4. 6】 Given $X \sim N(1,4)$, determine the probabilities $P\{0<X<1.5\}$.

Solution:

From Theorem 4. 1, we would calculate the CDF of X via standardizing and using the $\Phi(z)$ function:

$$P\{X\leqslant x\}=P\{\frac{X-\mu}{\sigma}\leqslant\frac{x-\mu}{\sigma}\}=\Phi\left(\frac{x-\mu}{\sigma}\right).$$

So, $P\{0<X<1.5\}=\Phi[(1.5-1)/2]-\Phi[(0-1)/2]=\Phi(0.25)-\Phi(-0.5)$

$=\Phi(0.25)-[1-\Phi(0.5)]=0.5987-1+0.6915=0.2902.$

【Example 4. 7】 Given $X \sim N(\mu, \sigma^2)$, determine the following probabilities $P\{|X-\mu|<\sigma\}, P\{|X-\mu|<2\sigma\}, P\{|X-\mu|<3\sigma\}$.

Solution: Standardize X, we get

$$P\left\{\left|\frac{X-\mu}{\sigma}\right|<1\right\}=P\left\{-1<\frac{X-\mu}{\sigma}<1\right\}=\Phi(1)-\Phi(-1)=2\Phi(1)-1=0.6826.$$

Similarly,

$$P\{|X-\mu|<2\sigma\}=2\Phi(2)-1=0.9544,$$
$$P\{|X-\mu|<3\sigma\}=2\Phi(3)-1=0.9974.$$

The results of Example 4. 7 reveal a very important statistical law, namely the so-called **"3σ principle"**[1] of normal distribution. It shows that in a random test, the normal random variables can almost certainly fall into $(\mu-3\sigma, \mu+3\sigma)$. The probability outside this interval is less than 3/1000, it is a small probability event, in practice, it is often considered that it is impossible to appear. In enterprise management, people often apply "3σ principle" to quality control. If a quality index of the product exceeds the limited range of 3σ, then the production process must be out of order and must be checked for adjustment.

---

[1] 3σ principle: 3σ 原则

4.2 Multiple Continuous Random Variables

4.2.1 Joint Distribution

There are many experimental situations in which more than one random variable will be of interest to an investigator. Now we will extend the notion of a PDF to the case of multiple random variables. In complete analogy with discrete random variables, we introduce joint, marginal, and conditional PDFs. Their intuitive interpretation as well as their main properties parallel the discrete case.

Definition 4.7 We say that two continuous random variables associated with a common experiment are jointly continuous and can be described in terms of a joint PDF $f_{X,Y}$, if $f_{X,Y}$ is a nonnegative function that satisfies

$$P\{(X,Y) \in G\} = \iint_{(x,y)\in G} f_{X,Y}(x,y)\mathrm{d}x\mathrm{d}y. \tag{4.13}$$

As in the case of one random variable, the advantage of working with the CDF is that it applies equally well to discrete and continuous random variables.

Proposition 4.7 If X and Y are described by a joint PDF $f_{X,Y}$, then joint CDF is

$$F_{X,Y}(x,y) = P\{X \leqslant x, Y \leqslant y\} = \int_{-\infty}^{x}\int_{-\infty}^{y} f_{X,Y}(s,t)\mathrm{d}s\mathrm{d}t. \tag{4.14}$$

Conversely, the PDF can be got from the CDF by differentiating:

$$f_{X,Y}(x,y) = \frac{\partial^2 F_{X,Y}(x,y)}{\partial x \partial y}. \tag{4.15}$$

In the particular case where G is a rectangle of the form $G=[a,b]\times[c,d]$, we have

$$P\{a \leqslant X \leqslant b, c \leqslant Y \leqslant d\} = \int_{c}^{d}\int_{a}^{b} f_{X,Y}(s,t)\mathrm{d}s\mathrm{d}t. \tag{4.16}$$

Furthermore, by letting G be the entire two-dimensional plane, we obtain the normalization property

$$\int_{-\infty}^{+\infty}\int_{-\infty}^{+\infty} f_{X,Y}(x,y)\mathrm{d}x\mathrm{d}y = 1. \tag{4.17}$$

【Example 4.8】 (**Two-Dimensional Uniform PDF**) Jack and Rose have a date at a given time, and each will arrive at the meeting place with a delay between 0 and 1 hour. Let X and Y denote the delays of Jack and Rose, respectively. Assuming that no pairs (X,Y) in the square $[0,1]\times[0,1]$ are more likely than others, a natural model involves a joint PDF of the form

$$f_{X,Y}(x,y) = \begin{cases} c, & 0\leqslant x\leqslant 1, 0\leqslant y\leqslant 1, \\ 0, & \text{otherwise}. \end{cases}$$

Determine the constant c.

Solution: For this PDF to satisfy the normalization property

$$\int_{-\infty}^{+\infty}\int_{-\infty}^{+\infty} f_{X,Y}(x,y)\,\mathrm{d}x\,\mathrm{d}y=\int_{0}^{1}\int_{0}^{1} c\,\mathrm{d}x\,\mathrm{d}y=1.$$

we must have $c=1$.

This is an example of a uniform PDF on the unit square. We can give more general definition of two-dimensional uniform distribution.

Definition 4.8 Let's fix some subset G of the two-dimensional plane. (X,Y) is said to follow **two-dimensional uniform distribution**[1] on G if its joint PDF is

$$f_{X,Y}(x,y)=\begin{cases}\dfrac{1}{\text{area of } G}, & (x,y)\in G,\\ 0, & \text{otherwise.}\end{cases} \tag{4.18}$$

Just as the most useful univariate distribution in statistical practice is the normal distribution, the most useful joint distribution for two random variables is the bivariate normal distribution, though its PDF is somewhat complicated.

Definition 4.9 A two-dimensional random variable (X,Y) is said to follow **bivariate normal distribution**[2] denoted by $N(\mu_1,\mu_2,\sigma_1^2,\sigma_2^2,\rho)$ if its joint PDF is

$$f(x,y)=\frac{1}{2\pi\sigma_1\sigma_2\sqrt{1-\rho^2}}\mathrm{e}^{-\frac{1}{2(1-\rho^2)}\left[\frac{(x-\mu_1)^2}{\sigma_1^2}-2\rho\frac{(x-\mu_1)(y-\mu_2)}{\sigma_1\sigma_2}+\frac{(y-\mu_2)^2}{\sigma_2^2}\right]},$$

$$(-\infty<x<+\infty,\ -\infty<y<+\infty) \tag{4.19}$$

A graph of this PDF, the density surface, appears in Figure 4.8, it's a curved surface with a high center and a low circumference.

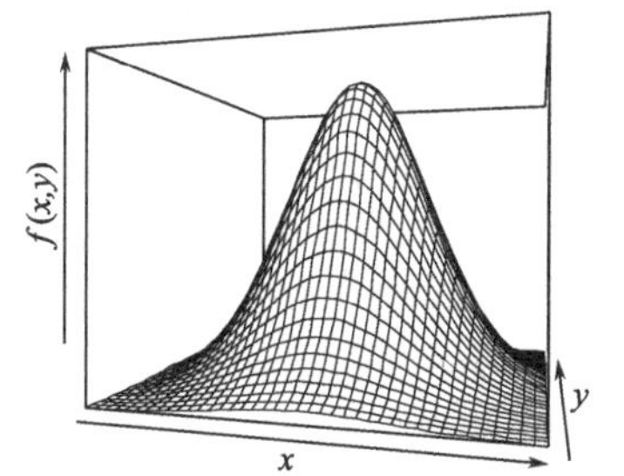

Figure 4.8 A graph of the bivariate normal PDF

It's not always convenient to calculate probabilities by integrating bivariate normal PDF, so we often employ numerical integration techniques for this purpose with help of software packages.

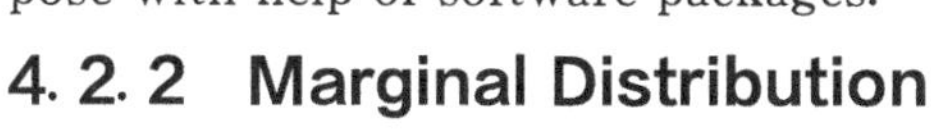

4.2.2 Marginal Distribution

The joint PDF contains all conceivable probabilistic information on the random variables X and Y, including the marginal distributions of two variables and the dependencies between X and Y. All in all, it allows us to calculate the probability of any event that can

---

[1] two-dimensional uniform distribution：二维均匀分布

[2] bivariate normal distribution：二维正态分布

be defined in terms of these two random variables.

As a special case, joint PDF can be used to calculate the probability of an event involving only one of them, such as event $\{X \leqslant x\}$. It's clear that event $\{X \leqslant x\}$ is equivalent to event $\{X \leqslant x, -\infty < Y < +\infty\}$, so we have

$$P\{X \leqslant x\} = P\{X \leqslant x, -\infty < Y < +\infty\}.$$

Express it in integral form,

$$\int_{-\infty}^{x} f_X(s)\mathrm{d}s = \int_{-\infty}^{+\infty}\int_{-\infty}^{x} f_{X,Y}(s,t)\mathrm{d}s\mathrm{d}t = \int_{-\infty}^{x}\left[\int_{-\infty}^{+\infty} f_{X,Y}(s,t)\mathrm{d}t\right]\mathrm{d}s$$

By taking the derivative of x on both sides, we find that the marginal PDF of X is given by

$$f_X(x) = \int_{-\infty}^{+\infty} f_{X,Y}(x,y)\mathrm{d}y \tag{4.20}$$

Similarly,

$$f_Y(y) = \int_{-\infty}^{+\infty} f_{X,Y}(x,y)\mathrm{d}x \tag{4.21}$$

【Example 4.9】 Prove that **the marginal distribution of two-dimensional normal distribution is still normal distribution.** Specifically, if two-dimensional random variable $(X,Y) \sim N(\mu_1,\mu_2,\sigma_1^2,\sigma_2^2,\rho)$, then $X \sim N(\mu_1,\sigma_1^2), Y \sim N(\mu_2,\sigma_2^2)$.

Proof: Since $(X,Y) \sim N(\mu_1,\mu_2,\sigma_1^2,\sigma_2^2,\rho)$, we can write the joint PDF

$$f(x,y) = \frac{1}{2\pi\sigma_1\sigma_2\sqrt{1-\rho^2}} \mathrm{e}^{-\frac{1}{2(1-\rho^2)}\left[\frac{(x-\mu_1)^2}{\sigma_1^2} - 2\rho\frac{(x-\mu_1)(y-\mu_2)}{\sigma_1\sigma_2} + \frac{(y-\mu_2)^2}{\sigma_2^2}\right]}.$$

Then,

$$\begin{aligned} f_X(x) &= \int_{-\infty}^{+\infty} f_{X,Y}(x,y)\mathrm{d}y \\ &= \int_{-\infty}^{+\infty} \frac{1}{2\pi\sigma_1\sigma_2\sqrt{1-\rho^2}} \mathrm{e}^{-\frac{1}{2(1-\rho^2)}\left[\frac{(x-\mu_1)^2}{\sigma_1^2} - 2\rho\frac{(x-\mu_1)(y-\mu_2)}{\sigma_1\sigma_2} + \frac{(y-\mu_2)^2}{\sigma_2^2}\right]}\mathrm{d}y. \end{aligned}$$

Let $\frac{x-\mu_1}{\sigma_1} = u, \frac{y-\mu_2}{\sigma_2} = v$, so that $f_X(x)$ can be reduced to

$$\begin{aligned} f_X(x) &= \frac{1}{2\pi\sigma_1\sqrt{1-\rho^2}} \int_{-\infty}^{+\infty} \exp\left[-\frac{1}{2(1-\rho^2)}(u^2 - 2\rho uv + v^2)\right]\mathrm{d}v \\ &= \frac{1}{\sqrt{2\pi}\sigma_1}\mathrm{e}^{-\frac{u^2}{2}} \frac{1}{\sqrt{2\pi(1-\rho^2)}}\int_{-\infty}^{+\infty} \mathrm{e}^{-\frac{(\rho u - v)^2}{2(1-\rho^2)}}\mathrm{d}v \\ &= \frac{1}{\sqrt{2\pi}\sigma_1}\mathrm{e}^{-\frac{u^2}{2}} \end{aligned}$$

$$=\frac{1}{\sqrt{2\pi}\sigma_1}e^{-\frac{(x-\mu_1)^2}{2\sigma_1^2}}.$$

According to the form of PDF of normal distribution, we know that $X\sim N(\mu_1,\sigma_1^2)$.

In a similar way, we find out that $Y\sim N(\mu_2,\sigma_2^2)$.

4.2.3 Conditional Distribution

In many situations, we have to model continuous random variable Y based on an underlying phenomenon, represented by a random variable X with PDF $f_X(x)$. For example, suppose X and Y denote the lifetimes of the front and rear tires on a motorcycle, and it happens that $X=10,000$ miles, what now is the probability that Y is at most 15,000 miles? Questions of this sort can be answered by studying conditional probability distributions. This setting is similar to problems we discussed in the previous chapter. The only difference is that we are now dealing with continuous random variables.

In order to solve this problem, we have to define the conditional CDF to obtain the conditional PDF of the continuous random variable by finding the derivative.

Definition 4.10 For a given y, if the inequality $P\{y-\varepsilon<Y<y+\varepsilon\}>0$ holds for all fixed positive number ε, and for any real number x, the following limit exists,

$$\lim_{\varepsilon\to0^+}P\{\xi\leqslant x\mid y-\varepsilon<\eta\leqslant y+\varepsilon\}=\lim_{\varepsilon\to0^+}\frac{P\{\xi\leqslant x,y-\varepsilon<\eta\leqslant y+\varepsilon\}}{P\{y-\varepsilon<\eta\leqslant y+\varepsilon\}},$$

then this limit is called **conditional CDF**[1] under condition "$Y=y$", denoted by

$$P\{X\leqslant x\mid Y=y\}\text{ or }F_{X|Y}(x\mid y).$$

In fact,

$$F_{X|Y}(x\mid y)=\lim_{\varepsilon\to0^+}\frac{P\{X\leqslant x,y-\varepsilon<Y\leqslant y+\varepsilon\}}{P\{y-\varepsilon<Y\leqslant y+\varepsilon\}}=\lim_{\varepsilon\to0^+}\frac{\int_{-\infty}^{x}\int_{y-\varepsilon}^{y+\varepsilon}f(s,t)\,\mathrm{d}s\,\mathrm{d}t}{\int_{y-\varepsilon}^{y+\varepsilon}f_Y(t)\,\mathrm{d}t}.$$

According to the mean value theorem of definite integral, we have

$$F_{X|Y}(x\mid y)=\lim_{\varepsilon\to0^+}\frac{\int_{-\infty}^{x}f(s,\eta_1)2\varepsilon\,\mathrm{d}s}{f_Y(\eta_2)2\varepsilon}=\lim_{\varepsilon\to0^+}\int_{-\infty}^{x}\frac{f(s,\eta_1)}{f_Y(\eta_2)}\mathrm{d}s=\int_{-\infty}^{x}\frac{f(s,y)}{f_Y(y)}\mathrm{d}s,$$

where $\eta_1,\eta_2\in(y-\varepsilon,y+\varepsilon)$.

Take the derivative of x on both sides, we find that the conditional PDF of X

---

[1] conditional CDF：条件 CDF

conditional on Y is as follows,

$$f_{X|Y}(x|y)=\frac{f(x,y)}{f_Y(y)}. \tag{4.22}$$

Similarly,

$$f_{Y|X}(y|x)=\frac{f(x,y)}{f_X(x)}. \tag{4.23}$$

From above, we find that the joint PDF can be obtained from marginal PDF and conditional PDF,

$$f(x,y)=f_X(x)f_{Y|X}(y|x)=f_Y(y)f_{X|Y}(x|y). \tag{4.24}$$

【Example 4.10】 Let X take a random value in the interval (0, 1). When $X=x$ is observed, where $0<x<1$, Y takes a random value on the interval $(x,1)$, To find the PDF of Y.

Solution: From the question, we know that $X\sim U(0,1), Y\mid X=x\sim U(x,1)$, that is

$$f_X(x)=\begin{cases}1, & 0<x<1,\\ 0, & \text{otherwise.}\end{cases} \qquad f_{Y|X}(y|x)=\begin{cases}1/(1-x), & x<y<1,\\ 0, & \text{otherwise.}\end{cases}$$

So the joint PDF is

$$f(x,y)=f_X(x)f_{Y|X}(y|x)=\begin{cases}1/(1-x), & 0<x<y<1,\\ 0, & \text{otherwise.}\end{cases}$$

Then the marginal PDF is obtained by integrating the joint PDF,

$$f_Y(y)=\int_{-\infty}^{+\infty}f(x,y)\mathrm{d}x=\begin{cases}\int_0^y 1/(1-x)\mathrm{d}x=-\ln(1-y), & 0<y<1,\\ 0, & \text{otherwise.}\end{cases}$$

4.2.4 Independence of Continuous Random Variables

Similar to analogy with the discrete case, if the value of continuous random variable Y does not affect the probability distribution of the continuous random variable X, then X is independent of Y, i.e. $f_{X|Y}(x\mid y)=\frac{f(x,y)}{f_Y(y)}=f_X(x)$ or $f_{X,Y}(x,y)=f_X(x)f_Y(y)$.

Lemma 4.1 Two continuous random variables X and Y are independent if their joint PDF is the product of the marginal PDFs:

$$f_{X,Y}(x,y)=f_X(x)f_Y(y) \quad \text{for all } x,\ y. \tag{4.25}$$

【Example 4.11】 Suppose two-dimensional random $(X,Y)\sim N(\mu_1,\mu_2,\sigma_1^2,\sigma_2^2,\rho)$. Prove that X and Y are independent if and only if the parameter $\rho=0$.

Proof:

"Proof of sufficiency"

We write the joint PDF as

$$f(x,y)=\frac{1}{2\pi\sigma_1\sigma_2\sqrt{1-\rho^2}}e^{-\frac{1}{2(1-\rho^2)}\left[\frac{(x-\mu_1)^2}{\sigma_1^2}-2\rho\frac{(x-\mu_1)(y-\mu_2)}{\sigma_1\sigma_2}+\frac{(y-\mu_2)^2}{\sigma_2^2}\right]}$$

If $\rho=0$, then the joint PDF is reduced to

$$\begin{aligned}f(x,y)&=\frac{1}{2\pi\sigma_1\sigma_2}e^{-\frac{1}{2}\left[\frac{(x-\mu_1)^2}{\sigma_1^2}+\frac{(y-\mu_2)^2}{\sigma_2^2}\right]}\\&=\frac{1}{\sqrt{2\pi}\sigma_1}e^{\frac{(x-\mu_1)^2}{-2\sigma_1^2}}\frac{1}{\sqrt{2\pi}\sigma_2}e^{\frac{(y-\mu_2)^2}{-2\sigma_2^2}}\\&=f_X(x)f_Y(y),\end{aligned}$$

which means that X and Y are independent.

"Proof of necessity"

If X and Y are independent, then

$$f_{X,Y}(x,y)=f_X(x)f_Y(y),$$

i. e.

$$\frac{1}{2\pi\sigma_1\sigma_2\sqrt{1-\rho^2}}e^{-\frac{1}{2(1-\rho^2)}\left[\frac{(x-\mu_1)^2}{\sigma_1^2}-2\rho\frac{(x-\mu_1)(y-\mu_2)}{\sigma_1\sigma_2}+\frac{(y-\mu_2)^2}{\sigma_2^2}\right]}=\frac{1}{\sqrt{2\pi}\sigma_1}e^{\frac{(x-\mu_1)^2}{-2\sigma_1^2}}\frac{1}{\sqrt{2\pi}\sigma_2}e^{\frac{(y-\mu_2)^2}{-2\sigma_2^2}}.$$

Substitute $x=\mu_1$ and $y=\mu_2$ into the upper formula, we get

$$\frac{1}{2\pi\sigma_1\sigma_2\sqrt{1-\rho^2}}=\frac{1}{2\pi\sigma_1\sigma_2}.$$

That's to say $\rho=0$.

Therefore, to judge whether two random variables in a two-dimensional normal distribution are independent, we only need to see whether ρ is equal to zero or not.

4.3 Derived Distributions of Continuous Variable

Any function $Z=g(X,Y)$ of a continuous random variable (X,Y) is still a continuous random variable when $g(\cdot)$ is a continuous function. In some cases, we may be interested in the probability distribution of Z.

When the joint PDF $f_{X,Y}(x,y)$ of (X,Y) is known, the following two-step approach can be used:

Firstly, calculate the CDF of Z

$$F_Z(z)=P\{g(X,Y)\leqslant z\}=\iint_{G=\{(x,y)\mid g(x,y)\leqslant z\}} f_{X,Y}(x,y)\mathrm{d}x\mathrm{d}y. \quad (4.26)$$

Secondly, differentiate to obtain the PDF of Z:

$$f_Z(z)=\frac{\mathrm{d}F_Z(z)}{\mathrm{d}z}.$$

The two-step procedure that first calculates the CDF and then differentiates to obtain the PDF also applies to functions of one random variable.

In particular, if $g(\cdot)$ is a function of one variable, suppose that $Y=g(X)$, when the PDF $f_X(x)$ of a continuous random variable X is known, the following two-step approach can be used: Firstly, calculate the CDF of Y:

$$F_Y(y)=P\{g(X)\leqslant y\}=\int_{\{x\mid g(x)\leqslant y\}}^{+\infty} f_X(x)\mathrm{d}x. \quad (4.27)$$

Secondly, differentiate to obtain the PDF of Y:

$$f_Y(y)=\frac{\mathrm{d}F_Y(y)}{\mathrm{d}y}.$$

【Example 4.12】 Let $Y=g(X)=X^2$, where X is a random variable with known PDF. For any $y\geqslant 0$, we have

$$F_Y(y)=P\{Y\leqslant y\}=P\{X^2\leqslant y\}$$
$$=P\{-\sqrt{y}\leqslant X\leqslant\sqrt{y}\}=F_X(\sqrt{y})-F_X(-\sqrt{y}).$$

Therefore, by differentiating and using the chain rule,

$$f_X(y)=\frac{1}{2\sqrt{y}}f_X(\sqrt{y})+\frac{1}{2\sqrt{y}}f_X(-\sqrt{y}),y\geqslant 0.$$

If g is a strictly monotonic function, then we will have the following conclusion.

Theorem 4.2 **(PDF formula for a monotonic function of a continuous random variable)** Suppose that $g(\cdot)$ is a monotonic and derivable function defined in a certain interval I, and $h(\cdot)$ is the invert function of $g(\cdot)$, i.e. $y=g(x),x\in I\Leftrightarrow x=g^{-1}(y)=h(y)$. Then the PDF of Y is given by

$$f_Y(y)=f_X(h(y))|h'(y)|. \quad (4.28)$$

Proof: When $g(\cdot)$ is monotonically increasing function, then event "$g(X)\leqslant y$" is equivalent to event "$X\leqslant h(y)$", so

$$F_Y(y)=P\{g(X)\leqslant y\}=P\{X\leqslant h(y)\}=\int_{-\infty}^{h(y)} f_X(x)\mathrm{d}x.$$

Derivative with respect to y,

$$f_Y(y)=\frac{\mathrm{d}[F_Y(y)]}{\mathrm{d}y}=f_X(x)h'(y),$$

where $h'(y)>0$. Because $g(\cdot)$ is monotonically increasing, then $h(\cdot)$ is monotonically increasing so its derivative is positive.

If $g(\cdot)$ is monotonically increasing function, then event "$g(X) \leqslant y$" is equivalent to event "$X \geqslant h(y)$", so

$$F_Y(y)=P\{g(X)\leqslant y\}=P\{X\geqslant h(y)\}=1-P\{X\leqslant h(y)\}=1-F[h(y)].$$

Derivative with respect to y using chain rule:

$$f_Y(y)=\frac{\mathrm{d}[F_Y(y)]}{\mathrm{d}y}=-f_X(h(y))h'(y)=f_X(h(y))|h'(y)|,$$

where $h'(y)<0$, because $h(\cdot)$ is monotone decreasing like $g(\cdot)$, so its derivative is negative.

【Example 4.13】 (A linear function of a normal random variable is normal)
Suppose that X is a normal random variable with mean μ and variance σ^2, and let $Y=aX+b$, where a and b are some scalars. We have

$$f_X(x)=\frac{1}{\sqrt{2\pi}\sigma}\mathrm{e}^{-(x-\mu)^2/2\sigma^2}.$$

Therefore,

$$f_Y(y)=\frac{1}{|a|}f_X\left(\frac{y-b}{a}\right)=\frac{1}{|a|}\frac{1}{\sqrt{2\pi}\sigma}\mathrm{e}^{-\left(\frac{y-b}{a}-\mu\right)^2/2\sigma^2}$$

$$=\frac{1}{\sqrt{2\pi}|a|\sigma}\mathrm{e}^{-(y-b-a\mu)^2/2a^2\sigma^2}.$$

We recognize this as a normal PDF with mean $a\mu+b$ and variance $a^2\sigma^2$, i.e. $Y=aX+b\sim N(a\mu+b, a^2\sigma^2)$, where $X\sim N(\mu,\sigma^2)$.

Theorem 4.3 **(Derived distribution of sum function)** Let the PDF of the two-dimensional random variable (X,Y) be $f(x,y)$, then the PDF of the random variable $Z=X+Y$ can be represented by the following **convolution formula**[1]:

$$f_Z(z)=\int_{-\infty}^{+\infty}f(x,z-x)\mathrm{d}x \text{ or } f_Z(z)=\int_{-\infty}^{+\infty}f(z-y,y)\mathrm{d}y. \tag{4.29}$$

In particular, when X and Y are independent of each other, we have

$$f_Z(z)=\int_{-\infty}^{+\infty}f_X(x)f_Y(z-x)\mathrm{d}x \text{ or } f_Z(z)=\int_{-\infty}^{+\infty}f_X(z-y)f_Y(y)\mathrm{d}y. \tag{4.30}$$

Proof:

First of all, find the CDF of Z

$$F_Z(z)=\iint\limits_{G:x+y\leqslant z}f(x,y)\mathrm{d}x\mathrm{d}y=\int_{-\infty}^{+\infty}\left[\int_{-\infty}^{z-x}f(x,y)\mathrm{d}y\right]\mathrm{d}x.$$

Take the variable substitution $y=u-x$, so

$$F_Z(z)=\int_{-\infty}^{+\infty}\left[\int_{-\infty}^{z}f(x,u-x)\mathrm{d}u\right]\mathrm{d}x=\int_{-\infty}^{z}\left[\int_{-\infty}^{+\infty}f(x,u-x)\mathrm{d}x\right]\mathrm{d}u.$$

---

[1] convolution formula：卷积公式

Derivative with respect to z, we get PDF of Z,

$$f_Z(z)=\int_{-\infty}^{+\infty} f(x,z-x)\mathrm{d}x.$$

Obviously, when X and Y are independent,

$$f_Z(z)=\int_{-\infty}^{+\infty} f_X(x)f_Y(z-x)\mathrm{d}x.$$

According to the symmetry, we can get

$$f_Z(z)=\int_{-\infty}^{+\infty} f(z-y,y)\mathrm{d}y,$$

and

$$f_Z(z)=\int_{-\infty}^{+\infty} f_X(z-y)f_Y(y)\mathrm{d}y,$$

when X and Y are independent each other.

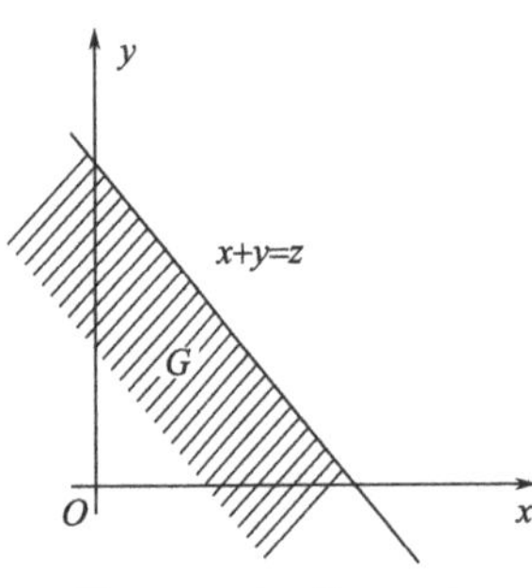

Figure 4.9 Integral region diagram

Theorem 4.4 (**Derived distribution of maximum and minimum function**) Let X and Y be independent random variables and their distribution functions are recorded as $f_X(x)$ and $f_Y(y)$ respectively. Let $M=\max(X,Y)$ and $N=\min(X,Y)$, then the CDFs of M and N are as follows:

$$F_M(z)=F_X(z)F_Y(z) \text{ and}$$
$$F_N(z)=1-[1-F_X(z)][1-F_Y(z)]. \quad (4.31)$$

Proof: When $M=\max(X,Y)$, the event "$M\leqslant z$" is equal to event "$X\leqslant z, Y\leqslant z$", therefore

$$F_M(z)=P\{M\leqslant z\}=P\{X\leqslant z,Y\leqslant z\}=P\{X\leqslant z\}P\{Y\leqslant z\}=F_X(z)F_Y(z).$$

Similarly, when $N=\min(X,Y)$, the event "$N>z$" is equal to event "$X>z,Y>z$", therefore

$$\begin{aligned}F_N(z)&=P\{N\leqslant z\}=1-P\{N>z\}\\&=1-P\{X>z,Y>z\}=1-P\{X>z\}P\{Y>z\}\\&=1-[1-F_X(z)][1-F_Y(z)].\end{aligned}$$

【Example 4.14】 (**Derived distribution of maximum function**) Suppose a circuit system L is composed of two independent subsystems L_1 and L_2, which are connected in (a) Series, (b) Parallel and (c) Standby (when L_1 is damaged, L_2 begins to work). As is shown in Figure 4.10. X and Y are the lifetimes of L_1 and L_2 respectively, which obey the exponential distribution with parameters λ and u ($\lambda>0, u>0, \lambda\neq u$). Try to find out the distribution of the lifetime of the system L according to the above three connection modes.

Solution:

(a) Series system

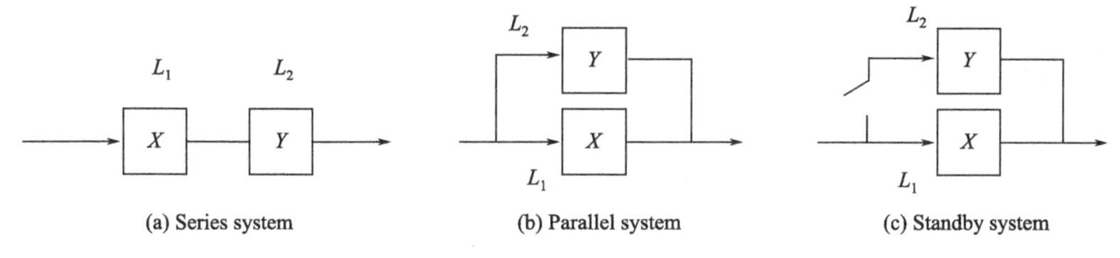

Figure 4. 10　A circuit system

When there is a damage in L_1 **or** L_2, the system L stops working, so the life of L is $Z=\min(X,Y)$ in which X and Y are independent each other, then the CDF of Z is

$$F_Z(z)=1-[1-F_X(z)][1-F_Y(z)]$$

$$=\begin{cases}1-e^{-(\lambda+u)z}, & z>0,\\ 0, & z\leqslant 0.\end{cases}$$

Derivative with respect to z, we get PDF of Z

$$f_Z(z)=\begin{cases}(\lambda+u)e^{-(\lambda+u)z}, & z>0,\\ 0, & z\leqslant 0.\end{cases}$$

(b) Parallel system

When there are damages in both L_1 **and** L_2, the system L stops working, so the life of L is $Z=\max(X,Y)$ in which X and Y are independent each other, then the CDF of Z is

$$F_Z(z)=F_X(z)F_Y(z)=\begin{cases}(1-e^{-\lambda z})(1-e^{-uz}), & z>0,\\ 0, & z\leqslant 0.\end{cases}$$

Derivative with respect to z, we get PDF of Z

$$f_Z(z)=\begin{cases}\lambda e^{-\lambda z}+ue^{-uz}-(\lambda+u)e^{-(\lambda+u)z}, & z>0,\\ 0, & z\leqslant 0.\end{cases}$$

(c) Standby system

When L_1 is damaged, the system L_2 begins to work, so the lifetime Z of the whole system L is the sum of the lifetime of L_1 and L_2, i. e. $Z=X+Y$, in which X and Y are independent each other, then the PDF of Z is

$$f_Z(z)=\int_{-\infty}^{+\infty}f_X(x)f_Y(z-x)\mathrm{d}x$$

$$=\begin{cases}\lambda e^{-\lambda x}\cdot\mu e^{-\mu(z-x)}, & x>0, z-x>0,\\ 0, & \text{otherwise}\end{cases}$$

$$=\begin{cases}\dfrac{\lambda u}{u-\lambda}(e^{-\lambda z}-e^{-uz}), & z>0,\\ 0, & z\leqslant 0.\end{cases}$$

Exercises

1. The current in a certain circuit as measured by an ammeter is a continuous random variable X with the following density function:

$$f(x)=\begin{cases}0.75x+0.2, & 3\leqslant x\leqslant 5,\\ 0, & \text{otherwise.}\end{cases}$$

(1) Graph the PDF and verify that the total area under the density curve is indeed 1.

(2) Calculate $P\{X\leqslant 4\}$. How does this probability compare to $P\{X<4\}$?

(3) Calculate $P\{3.5\leqslant X\leqslant 4.5\}$ and also $P\{X>4.5\}$.

2. A college professor never finishes his lecture before the end of the hour and always finishes his lectures within 2min after the hour. Let $X=$ the time that elapses between the end of the hour and the end of the lecture and suppose the PDF of X is

$$f(x)=\begin{cases}kx^2, & 0\leqslant x\leqslant 2,\\ 0, & \text{otherwise.}\end{cases}$$

(1) Find the value of k. [Hint: Total area under the graph of $f(x)$ is 1.]

(2) What is the probability that the lecture ends within 1 min of the end of the hour?

(3) What is the probability that the lecture continues beyond the hour for between 60 and 90 sec?

(4) What is the probability that the lecture continues for at least 90 sec beyond the end of the hour?

3. Let Z be a random variable with probability density function

$$f_Z(z)=\begin{cases}-\dfrac{3}{2}z^2+\dfrac{3}{2}, & 0\leqslant z\leqslant 1,\\ 0, & \text{otherwise.}\end{cases}$$

(1) Tabulate the cumulative distribution function of Z.

(2) Give a check to show that your cumulative distribution function in part (1) is correct.

(3) Calculate $P\{Z>0.5\}$.

4. Let Y be a random variable with PDF

$$f_Y(y)=\mathrm{e}^{-2|y|}, y\in R.$$

(1) Calculate $P\{-1<Y<2\}$.

(2) If it is known that $P\{Y>0\}$, calculate $P\{Y<1|Y>0\}$. (Note that this is

not the same as $P\{0<Y<1\}$.)

5. Someone woke up in the morning and found that the watch had stopped. he turned on the radio and waited for the hourly time, assuming that it was possible for him to wake up at any time before the radio announces the hour next time.

   (1) Try to find the PDF of the waiting time X (minutes).

   (2) Try to find the probability that the waiting time is less than 5 minutes.

6. Let X = the time between two successive arrivals at the drive-up window of a local bank. If X has an exponential distribution with $\lambda=1$, compute the following probabilities of

   (1) $P\{X\leqslant 4\}$; (2) $P\{2\leqslant X\leqslant 5\}$.

7. Let Z be a standard normal random variable and calculate the following probabilities.

   (1) $P\{0\leqslant Z\leqslant 2.17\}$; (2) $P\{0\leqslant Z\leqslant 1\}$; (3) $P\{-2.50\leqslant Z\leqslant 0\}$;

   (4) $P\{Z\leqslant 1.37\}$; (5) $P\{-2.50\leqslant Z\leqslant 2.50\}$; (6) $P\{Z\geqslant -1.75\}$;

   (7) $P\{-1.50\leqslant Z\leqslant 2\}$; (8) $P\{1.37\leqslant Z\leqslant 2.50\}$; (9) $P\{Z\geqslant 1.50\}$;

   (10) $P\{|Z|\leqslant 2.50\}$.

8. In each case, determine the value of the constant c that makes the probability statement correct.

   (1) $\Phi(c)=0.9838$; (2) $P\{0\leqslant Z\leqslant c\}=0.291$; (3) $P\{c\leqslant Z\}=1.21$;

   (4) $P\{-c\leqslant Z\leqslant c\}=0.668$; (5) $P\{|Z|\geqslant c\}=0.016$.

9. Suppose the force acting on a column that helps to support a building is a normally distributed random variable X, $X\sim N(15,1.25^2)$. Compute the following probabilities.

   (1) $P\{X\leqslant 15\}$; (2) $P\{X\leqslant 17.5\}$; (3) $P\{X\geqslant 10\}$;

   (4) $P\{14\leqslant X\leqslant 18\}$; (5) $P\{|X-15|\leqslant 3\}$.

10. A patient with gastroesophageal reflux disease is treated with a new drug to relieve pain from heartburn. Following treatment, the time until the patient next experiences the symptoms is recorded. The doctor treating the patient thinks there is a 50% chance that the patient will stay symptom-free for at least 30 days.

    (1) If the time until recurrence of symptoms is to be modelled using an exponential distribution, find the rate parameter of this distribution based on the doctor's judgement .

    (2) What is the probability the patient will remain symptom-free for at least 60 days?

    (3) If, after 30 days, the patient has remained symptom-free, what is the

probability the patient will be symptom-free for at least another 30 days? (Note that the answer to part (2) is different).

11. Suppose $X\sim\text{Exp}(1)$. Without using a calculator, find the value of $P\{\ln 1\leqslant X\leqslant \ln 2\}$.

12. Let $U\sim U[0,1]$, let $Y=U^2$ and $Z=-\frac{1}{\lambda}\ln(1-U),\lambda>0$.

(1) Calculate $P\{Y\leqslant 0.09\}$;

(2) Derive an expression for $P\{Z\leqslant z\}$.

13. Let the random variable X follow the exponential distribution with parameter 2,

(1) What probability distribution does $Y=1-e^{-2X}$ follow?

(2) What probability distribution does $Y=X^2$ follow?

14. Each front tire on a particular type of vehicle is supposed to be filled to a pressure of 26 psi. Suppose the actual air pressure in each tire is a random variable—X for the right tire and Y for the left tire, with joint PDF

$$f(x,y)=\begin{cases}K(x^2+y^2), & 20\leqslant x\leqslant 30, 20\leqslant y\leqslant 30,\\ 0, & \text{otherwise.}\end{cases}$$

(1) What is the value of K?

(2) What is the probability that both tires are under filled?

(3) What is the probability that the difference in air pressure between the two tires is at most 2 psi?

(4) Determine the (marginal) distribution of air pressure in the right tire alone.

(5) Are X and Y independent random variables?

15. Jack and Rose have agreed to meet between 5: 00 p. m. and 6: 00 p. m. for dinner at a local health-food restaurant. Let X =Jack's arrival time and Y= Rose's arrival time. Suppose X and Y are independent with each uniformly distributed on the interval $[5,6]$.

(1) What is the joint PDF of X and Y?

(2) What is the probability that they both arrive between 5:15 and 5:45?

(3) If the first one to arrive will wait only 10 min before leaving to eat elsewhere, what is the probability that they have dinner at the health-food restaurant? [Hint: The event of interest is $A=\{(x,y)\mid |x-y|\leqslant 1/6\}$.]

16. Two components of a minicomputer have the following joint PDF for their useful lifetimes X and Y:

$$f(x,y)=\begin{cases}x e^{-x(1+y)}, & x\geqslant 0 \text{ and } y\geqslant 0,\\ 0, & \text{otherwise.}\end{cases}$$

(1) What is the probability that the lifetime X of the first component exceeds 3?

(2) What are the marginal PDF's of X and Y? Are the two lifetimes independent? Explain.

(3) What is the probability that the lifetime of at least one component exceeds 3?

17. Let the random variables X, Y be independent of each other and follow the standard normal distribution $N(0,1)$. The probability density function of $Z=X+Y$ is obtained.

18. Let the random variables X, Y be independent of each other and follow uniform distribution $U[-1,1]$. The probability density function of $Z=X+Y$ is obtained.

19. Let the output of an electronic device be measured five times, and the observed values are x_1, x_2, x_3, x_4, x_5. Let them be independent random variables and all follow the Rayleigh distribution with parameter $\sigma=2$, the PDF of which is $f(x)$,

$$f(x)=\begin{cases} \frac{x}{\sigma^2}e^{-\frac{x^2}{2\sigma^2}}, & z>0, \\ 0, & z<0, \end{cases}$$

where sigma is greater than zero. try to find:

(1) The CDF of $Y_1=\max(x_1, x_2, x_3, x_4, x_5)$;

(2) The probability of Y_1 is more than 4;

(3) The CDF of $Y_2=\min(x_1, x_2, x_3, x_4, x_5)$.

Chapter 5 Numerical Characteristics of Random Variables

The probability distribution gives a complete description of the uncertainty we have about a random variable X; It tells us how likely X takes a value or is in a certain range. However, there are other quantities that can tell us useful things about a random variable, which we can derive from the probability mass function. It would be desirable to summarize this information in a single representative number. We refer to the numbers that describe some characteristics of random variables as the **numerical characteristics**[1] of random variables. In this chapter, we will introduce several commonly used numerical characteristics of random variables: **expectation**[2], **variance**[3], **covariance**[4], **correlation coefficient**[5] and **moment**[6].

5.1 Expectation

5.1.1 Average & Expectation

In practice, the concept of "average" is often used. For example, in order to investigate the shooting ability of a sniper, n times tests have been carried out, the results of which are shown in the following table:

---

[1] numerical characteristics：数字特征

[2] expectation：数学期望

[3] variance：方差

[4] covariance：协方差

[5] correlation coefficient：相关系数

[6] moment：矩

score of shooting x_i	5	6	7	8	9	10
number of occurrences n_i	10	12	10	16	22	30

One of the scores of shooting x_i comes up with corresponding number of occurrences $n_i (i=1,2,3,4,5,6)$, in which $\sum_{i=1}^{6} n_i = n = 100$. So, the average score of this sniper is:

$$\frac{\sum_{i=1}^{6} n_i x_i}{n} = \sum_{i=1}^{6} x_i \frac{n_i}{n} = 5 \times \frac{10}{100} + 6 \times \frac{12}{100} + 7 \times \frac{10}{100} + 8 \times \frac{16}{100} + 9 \times \frac{22}{100} + 10 \times \frac{30}{100}$$
$$= 8.18. \tag{5.1}$$

Then what is the score that you "expect" the sniper to get next time? Is it 8.18? Maybe! However, if we carry out another n times experiments, maybe the average will become another number, because the frequency is random. Thus, the terms "expect" and "average" are a little ambiguous. In fact, if the number of experiments n is very large, it is reasonable to anticipate that the frequency $\frac{n_i}{n}$ is roughly equal to the probability of p_i: $\frac{n_i}{n} \approx p_i, i=1,2,3,4,5,6$.

Thus, if we put the probabilities instead of frequencies into equation (5.1), the score that you "expect" the sniper to get next time will be more reliable. Motivated by this example, we can introduce an important definition of expectation of discrete random variable, which can be obtained by summing $xp_X(x)$ over possible X values, it is the weighted average of x.

Definition 5.1 **(Expectation of discrete random variable)** The **expectation**[1] $E(X)$ of a discrete random variable X is defined as

$$E(X) = \sum_x x p_X(x). \tag{5.2}$$

Note:

- **Expectation** is also called **mean**[2], which reflects the centralized position of random variables.

As for continuous random variables, it is similar to the discrete case except that the PMF is replaced by the PDF, and summation is replaced by integration.

[1] expectation：期望

[2] mean：均值

Its mathematical properties are similar to the discrete case—after all, an integral is just a limiting form of a sum.

Definition 5.2 **(Expectation of continuous random variable)** The **expectation** $E(X)$ of a continuous random variable X with PDF $f(x)$ is

$$E(X)=\int_{-\infty}^{+\infty} xf(x)\mathrm{d}x. \tag{5.3}$$

Expected values often provide a convenient vehicle for choosing optimally between several candidate decisions that result in different expected rewards. If we view the expected reward of a decision as its "average payoff over a large number of trials," it is reasonable to choose a decision with maximum expected reward.

【Example 5.1】 Assume that two workers A and B produce the same product with equal daily production, with the number of defective products occurring in the day being X and Y, respectively, with PMF as shown in the table below:

X	0	1	2	3	4
p	0.4	0.3	0.2	0.1	0

Y	0	1	2	3	4
p	0.5	0.1	0.2	0.1	0.1

Try to compare the technical level of the two workers.

Solution: Calculate the expectations of X and Y from the PMF of them, we get

$$E(X)=0\times 0.4+1\times 0.3+2\times 0.2+3\times 0.1+4\times 0=1,$$
$$E(Y)=0\times 0.5+1\times 0.1+2\times 0.2+3\times 0.1+4\times 0=1.2.$$

Therefore, on average, B produces more defective products per day than A. in this sense, A's technology is better than B's.

【Example 5.2】 Suppose X is a continuous random variable with PDF as follows:

$$f(x)=\begin{cases}\dfrac{1}{\sqrt{1-x^2}}, & |x|<1,\\ 0, & |x|\geqslant 1.\end{cases}$$

Then, the expectation of X is

$$E(X)=\int_{-\infty}^{+\infty} xf(x)\mathrm{d}x=\int_{-1}^{1}\frac{x}{\pi\sqrt{1-x^2}}\mathrm{d}x=0.$$

Note:

- The expectation of a constant is the constant itself, $E(c)=c$.

5.1.2 Expectations for Functions of Random Variables

If X is a random variable, any real-valued function $Y=g(X)$ of X is also a random variable. If X is a discrete random variable with PMF known, Y must be a discrete random variable. Then, how to obtain the expectation of Y? Does it require the PMF of Y? We'll find later that it's unnecessary. Furthermore, if X

and Y are two discrete random variables, any multivariate real-valued function $Z=g(X,Y)$ of X and Y is also a random variable. We'll find later that it's also unnecessary to know the PMF of Z to obtain the expectation of Z. This method is based on the following rule.

Theorem 5.1 **(Expectations for Functions of Discrete Random Variables)**

(1) Let X be a random variable with PMF $p_X(x)$, and let $g(X)$ be a real valued function of X. Then, the expectation of the random variable $g(X)$ is given by

$$E[g(X)]=\sum_x g(x)p_X(x). \tag{5.4}$$

(2) Let (X,Y) be a two-dimensional random variable with joint PMF $p_{X,Y}(x,y)$, and let $g(X,Y)$ be a real valued function of (X,Y). Then, the expectation of the random variable $g(X,Y)$ is given by

$$E[g(X,Y)]=\sum_{x,y} g(x,y)p_{X,Y}(x,y). \tag{5.5}$$

To verify (1), we firstly use the formula $p_Y(y)=\sum_{\{x|g(x)=y\}} p_X(x)$ to calculate

$$E[g(X)]=E(y)=\sum_y yp_Y(y)=\sum_y y\sum_{\{x|g(x)=y\}} p_X(x)$$
$$=\sum_y \sum_{\{x|g(x)=y\}} g(x)p_X(x)=\sum_x g(x)p_X(x).$$

To verify (2), we use the formula $p_Z(z)=\sum_{\{(x,y)|g(x,y)=z\}} p_{X,Y}(x,y)$ to calculate

$$E[g(X,Y)]=E(Z)=\sum_z zp_Z(z)=\sum_z z\sum_{\{(x,y)|g(x,y)=z\}} p_{X,Y}(x,y)$$
$$=\sum_z \sum_{\{(x,y)|g(x,y)=z\}} g(x,y)p_{X,Y}(x,y)=\sum_{x,y} g(x,y)p_{X,Y}(x,y).$$

If X is a continuous random variable, Y can be a continuous random variable: for example, $Y=g(X)=X^2$. But Y can also turn out to be discrete. For example, suppose that $g(x)=1$ for $x>0$, and $g(x)=0$, otherwise. Then $Y=g(X)$ is a discrete random variable. The same case occurs on the function of multivariate continuous random variables. In each case, we can verify that the expectation for functions of continuous random variables in complete analogy with the discrete case.

Theorem 5.2 **(Expectations for Functions of Continuous Random Variables)**

(1) Let X be a random variable with PDF $f_X(x)$, and let $g(X)$ be a real valued function of X. Then, the expectation of the random variable $g(X)$ is given by

$$E[g(X)]=\int_{-\infty}^{+\infty} g(x)f_X(x)\mathrm{d}x. \tag{5.6}$$

(2) Let (X,Y) be a two-dimensional random variable with joint PDF $f_{X,Y}(x,y)$, and let $g(X,Y)$ be a real valued function of (X,Y). Then, the expectation of the random variable $g(X,Y)$ is given by

$$E[g(X,Y)]=\iint_{x,y} g(x,y)f_{X,Y}(x,y)\mathrm{d}x\mathrm{d}y. \tag{5.7}$$

【Example 5.3】 Let PMF of random variable X be

X	0	1	2
p	1/2	3/8	1/8

Then

$$E(X)=0\times\frac{1}{2}+1\times\frac{3}{8}+2\times\frac{1}{8}=\frac{5}{8},$$

$$E(X^2)=0^2\times\frac{1}{2}+1^2\times\frac{3}{8}+2^2\times\frac{1}{8}=\frac{7}{8},$$

$$E(3X+4)^2=(3\times0+4)^2\times\frac{1}{2}+(3\times1+4)^2\times\frac{3}{8}+(3\times2+4)^2\times\frac{1}{8}=\frac{311}{8}.$$

【Example 5.4】 Let the joint PMF of random variable (X,Y) be

X \ Y	0	1	2
0	4/16	4/16	1/16
1	4/16	2/16	0
2	1/16	0	0

Then

$$\begin{aligned}E(X+Y)&=\sum_{x,y}(x+y)p_{X,Y}(x,y)\\&=(0+0)\times\frac{4}{16}+(0+1)\times\frac{4}{16}+(0+2)\times\frac{1}{16}+(1+0)\times\frac{4}{16}+\\&\quad(1+1)\times\frac{2}{16}+(1+2)\times0+(2+0)\times\frac{1}{16}+(2+1)\times0+(2+2)\times0\\&=1.\end{aligned}$$

Proposition 5.1 If there exists the expectations of X and Y, then

$$E(cX)=cE(X);$$

$$E(X+Y)=E(X)+E(Y).$$

Furthermore, if X and Y are independent, then

$$E(XY)=E(X)E(Y).$$

Proof: We only prove the above conclusion for continuous random variables, and the discrete case can be proved similarly.

Suppose that the PDFs of X and Y are $f_X(x)$ and $f_Y(y)$, then

$$E(cX)=\int_{-\infty}^{+\infty} cxf_X(x)\mathrm{d}x=c\int_{-\infty}^{+\infty} xf_X(x)\mathrm{d}x=cE(X),$$

$$\begin{aligned}E(X+Y)&=\int_{-\infty}^{+\infty}\int_{-\infty}^{+\infty}(x+y)f_{X,Y}(x,y)\mathrm{d}x\mathrm{d}y\\&=\int_{-\infty}^{+\infty}\int_{-\infty}^{+\infty}xf_{X,Y}(x,y)\mathrm{d}x\mathrm{d}y+\int_{-\infty}^{+\infty}\int_{-\infty}^{+\infty}yf_{X,Y}(x,y)\mathrm{d}x\mathrm{d}y\\&=\int_{-\infty}^{+\infty}x\mathrm{d}x\int_{-\infty}^{+\infty}f_{X,Y}(x,y)\mathrm{d}y+\int_{-\infty}^{+\infty}y\mathrm{d}y\int_{-\infty}^{+\infty}f_{X,Y}(x,y)\mathrm{d}x\\&=\int_{-\infty}^{+\infty}xf_X(x)\mathrm{d}x+\int_{-\infty}^{+\infty}yf_Y(y)\mathrm{d}y\\&=E(X)+E(Y).\end{aligned}$$

The result shows that the expectation of the sum of random variables is the sum of the expectations of random variables.

When X and Y are independent, there must have $f_{X,Y}(x,y)=f_X(x)f_Y(y)$, thus

$$\begin{aligned}E(XY)&=\int_{-\infty}^{+\infty}\int_{-\infty}^{+\infty}xyf_{X,Y}(x,y)\mathrm{d}x\mathrm{d}y\\&=\int_{-\infty}^{+\infty}\int_{-\infty}^{+\infty}xyf_X(x)f_Y(y)\mathrm{d}x\mathrm{d}y\\&=\int_{-\infty}^{+\infty}xf_X(x)\mathrm{d}x\int_{-\infty}^{+\infty}yf_Y(y)\mathrm{d}y\\&=E(X)E(Y).\end{aligned}$$

Note:

- The more general conclusion can be drawn that the "expectation" of the linear combination of X and Y is the linear combination of their expectations. Since

$$E(aX+bY)=E(aX)+E(bY)=aE(X)+bE(Y).$$

- "Independence" is a sufficient condition for "the expectation of the product of random variables is equal to the product of their expectations ".

5.1.3 Moments of the Random Variable

As long as the probability distributions of a random variables are known, the expectation of any function of the random variables can be calculated (if it exists). Among them, the expectations of some special functions are defined as moments.

Definition 5.3 **(Moment of a random variable)**

For random variable X and Y, suppose that the expectations of the following functions of random variables exists.

The **k-th order origin moment**[1] of random variable X is defined as

$$E(X^k),(k=1,2,\cdots);$$

The **k-th order central moment**[2] of random variable X is defined as

$$E[(X-EX)^k],(k=1,2,\cdots);$$

The **$(k+l)$-th order mixing origin moment**[3] of random variable X and Y is defined as

$$E(X^kY^l),(k,l=1,2,\cdots);$$

The **$(k+l)$-th order mixing central moment**[4] of random variable X and Y is defined as

$$E[(X-EX)^k(Y-EY)^l],(k,l=1,2,\cdots).$$

For example, we define the 2nd moment of the random variable X as the expected value of the random variable X^2. More generally, we define the nth moment as $E(X^n)$, the expected value of the random variable X^n. With this terminology, the 1st moment of X is just the mean. Then what about other moments? We will talk about them later.

【Example 5.5】 Let the PDF of random variable X be

$$f_X(x)=\begin{cases}1, & 0\leqslant x\leqslant 1,\\ 0, & \text{others}.\end{cases}$$

Try to find the fifth order origin moment and the third-order central moment of X.

Solution: The fifth order origin moment of X is

$$E(X^5)=\int_{-\infty}^{+\infty}x^5 f_X(x)\mathrm{d}x=\int_0^1 x^5\cdot 1\mathrm{d}x=\frac{1}{6}.$$

The third-order central moment of X is

$$E(X-EX)^3=\int_{-\infty}^{+\infty}(x-EX)^3 f_X(x)\mathrm{d}x=\int_0^1\left(x-\frac{1}{6}\right)^3\cdot 1\mathrm{d}x=0.1203.$$

5.2 Variance

5.2.1 Variance & Standard Deviation

The expected value of X describes where the probability distribution is centered. In many practical problems, it is not enough to know the expected value.

---

[1] k-th order origin moment : k 阶原点矩

[2] k-th order central moment: k 阶中心矩

[3] $(k+l)$-th order mixing origin moment: $k+l$ 阶混合原点矩

[4] $(k+l)$-th order mixing central moment: $k+l$ 阶混合中心矩

For example, to compare the performance of two brands of watches whose PMFs of timing error X, Y are as follows:

X	-1	0	1
p	0.1	0.8	0.1

Y	-2	-1	0	1	2
p	0.1	0.2	0.4	0.2	0.1

It's easy to prove that the expected values of the two watches are equal, which are both zero, so it is impossible to compare the performance of the two watches in terms of the average timing error. But for values of X and Y we observe, X will be closer to $E(X)$ than Y will be to $E(Y)$, which shows that the timing error of the second watch is more scattered, not as stable as the first watch. So, in terms of dispersion, the first watch performs better. Here, in order to characterize the dispersion of the values of random variables, another numerical feature, **variance,** is introduced in addition to expectation. We use variance to describe how close a random variable is likely to be to its expected value.

Intuitively, we can express the dispersion degree of random variable by the average degree of the difference between random variables and its mean values, that is, $E[X-E(X)]$, but the truth is that the expected difference between a random variable and its mean is zero for any random variable X.

$$E[X-E(X)]=E(X)-E[E(X)]=E(X)-E(X)=0.$$

So this expression will not tell us anything useful about X. In fact, either X is larger or smaller than its mean, it shows deviation from the mean, and we should take the expectation of the absolute value of the deviation as an index to measure the dispersion degree, that is $E|X-E(X)|$, however, it's inconvenient to calculate. Therefore, we give the following definition to describe the dispersion degree of a random variable.

Definition 5.4 **(Variance of a random variable)**

The **variance**[1] of a random variable X is defined as

$$Var(X)=E[(X-E(X)]^2. \tag{5.8}$$

We denote the variance by $D(X)$ or σ_X^2.

Note:

- From the definition of variance, we find that it's in fact the second moment of random variable of X.

As the variance is defined as an expected squared difference, the variance will be expressed in units that are the square of the units of X. If we want a measure of

[1] variance：方差

spread that is in the same units as X, we take the square root of the variance.

Definition 5.5 **(Standard deviation of a random variable)**

The **standard deviation**[1] of a random variable X, denoted by $\sqrt{DX}$ or σ_X, is the square root of the variance of X,

$$\sigma_X=\sqrt{Var(X)}.$$

From the definition, variance can be considered as the expectation of the function $g(X)=[(X-E(X)]^2$ of the random variable X. Therefore,

- When X is a discrete random variable, $Var(X)=\sum_x[(x-E(X)]^2 p_X(x)$;
- When X is a continuous random variable,

$$Var(X)=\int_{-\infty}^{+\infty}[(x-E(X)]^2 f_X(x)\mathrm{d}x.$$

However, to compute the variance can be reduced by using an alternative formula.

Proposition 5.2 **(A shortcut formula for variance)**

$$Var(X)=E(X^2)-[E(X)]^2. \tag{5.9}$$

This expression means that variance can be calculated by the second origin moment and the expectation of X, which can be verified as follows:

According to the operational property of "expectation", it's easy to prove that

$$\begin{aligned}E(X-EX)^2 &=E[X^2-2XEX+(EX)^2]\\ &=EX^2-2(EX)^2+(EX)^2\\ &=EX^2-(EX)^2.\end{aligned}$$

Now let's recheck the performance of the two watches mentioned above.

【Example 5.6】 Suppose that the timing error of two brands of watches are X and Y whose PMFs are as follows:

X	-1	0	1
p	0.1	0.8	0.1

Y	-2	-1	0	1	2
p	0.1	0.2	0.4	0.2	0.1

Try to compare the performance of the two watches.

Solution: Firstly, calculate the expected values of X and Y,

$$E(X)=-1\times0.1+0\times0.8+1\times0.1=0;$$

$$E(Y)=-2\times0.1+(-1)\times0.2+0\times0.4+1\times0.2+2\times0.1=0.$$

Then, calculate the expected values of X^2 and Y^2,

[1] standard deviation：标准差

$$E(X^2)=(-1)^2\times 0.1+0^2\times 0.8+1^2\times 0.1=0.2;$$

$$E(Y^2)=(-2)^2\times 0.1+(-1)^2\times 0.2+0^2\times 0.4+1^2\times 0.2+2^2\times 0.1=1.2.$$

According to the shortcut formula for variance, we have

$$Var(X)=E(X^2)-(EX)^2=0.2;$$

$$Var(Y)=E(Y^2)-(EY)^2=1.2>Var(X).$$

So, from the point of view of travel time stability, the first brand of watch is due to the second one.

Proposition 5.3 If there exists the variances of X and Y, then:

(1) $Var(C)=0$;

(2) $Var(X+C)=Var(X)$;

(3) $Var(CX)=C^2Var(X)$;

(4) If X and Y are independent, $Var(X+Y)=Var(X)+Var(Y)$;

(5) $Var(X)=0$ if and only if X takes a constant with probability 1, i. e.

$$Var(X)=0\Leftrightarrow P\{X=C\}=1.$$

Proof:

(1) $Var(C)=E(C^2)-(EC)^2=C^2-C^2=0$;

(2) $Var(X+C)=E[(X+C)-E(X+C)]^2=E(X-EX)^2=Var(X)$;

(3) $Var(CX)=E[CX-E(CX)]^2=C^2E(X-EX)^2=C^2Var(X)$;

(4)
$$\begin{aligned}Var(X+Y)&=E[(X+Y)-E(X+Y)]^2\\&=E[(X-EX)+(Y-EY)]^2\\&=E(X-EX)^2+E(Y-EY)^2+2E[(X-EX)(Y-EY)]\\&=Var(X)+Var(Y)+2E[(X-EX)(Y-EY)],\end{aligned}\tag{5.10}$$

Consider the third item of formula (5.10),

$$\begin{aligned}E[(X-EX)(Y-EY)]&=E(XY\text{-}YEX-XEY+EXEY)\\&=E(XY)-EYEX-EXEY+EXEY\\&=E(XY)-EXEY.\end{aligned}$$

If X and Y are independent, there must have

$$E(XY)=EXEY,$$

that is to say,

$$E[(X-EX)(Y-EY)]=0.$$

Substitute it into formula (5.3), we obtain that

$$Var(X+Y)=Var(X)+Var(Y).$$

(5) $P\{X=C\}=1\Leftrightarrow\begin{cases}E(X)=C\times 1=C\\E(X^2)=C^2\times 1=C\end{cases}\Leftrightarrow Var(X)=E(X^2)-(EX)^2=0.$

【Example 5.7】 Suppose that X, Y and Z are independent random variables with expectations are 9, 20 and 12 respectively and variances are 2, 1 and 4 respectively. Random variable W is the linear function of X, Y and Z, where $W = X-2Y+5Z+1$. Try to obtain the expectation and variance of W.

According to the property of expectation and variance, we have

$$\begin{aligned} EW &= E(X-2Y+5Z+1) \\ &= EX-2EY+5EZ+1 \\ &= 9-2\times 20+5\times 12+1 \\ &= 30, \end{aligned}$$

$$\begin{aligned} Var(W) &= Var(X-2Y+5Z+1) \\ &= Var(X)+(-2)^2 Var(Y)+5^2 Var(Z)+Var(1) \\ &= 2+4\times 1+25\times 4+0 \\ &= 106. \end{aligned}$$

【Example 5.8】 Suppose that random variable X follows uniform distribution, $X\sim U[-0.5,0.5]$, Y is the function of X, $Y=\sin(\pi X)$, Try to get the variance of Y.

Solution: Since $X\sim U[-0.5,0.5]$, its PDF is $f(x)=\begin{cases}1, & -0.5<x<0.5, \\ 0, & \text{otherwise.}\end{cases}$

Then

$$EY=\int_{-\infty}^{+\infty}\sin\pi x\cdot f(x)\mathrm{d}x=\int_{-0.5}^{0.5}\sin\pi x\cdot 1\mathrm{d}x=0;$$

$$EY^2=\int_{-\infty}^{+\infty}\sin^2\pi x\cdot f(x)\mathrm{d}x=\int_{-0.5}^{0.5}\frac{1-\cos 2\pi x}{2}\cdot 1\mathrm{d}x=0.5;$$

Hence,

$$Var(Y)=EY^2-(EY)^2=0.5.$$

【Example 5.9】 Let (X, Y) be a two-dimensional random variable with joint PDF is

$$f(x,y)=\begin{cases}\dfrac{3xy}{16}, & (x,y)\in G, \\ 0, & (x,y)\notin G,\end{cases} \quad \text{in which} \quad G:\begin{cases}0\leqslant x\leqslant 2, \\ 0\leqslant y\leqslant x^2.\end{cases}$$

Try to get the variances of X and Y.

Solution: Though X is a random variable, it can also be seen as the function of X and Y. Therefore,

$$EX=\iint\limits_{(x,y)\in R^2} xf(x,y)\mathrm{d}x\mathrm{d}y=\iint\limits_{(x,y)\in G} x\,\frac{3xy}{16}\mathrm{d}x\mathrm{d}y=\int_0^2\mathrm{d}x\int_0^{x^2} x\,\frac{3xy}{16}\mathrm{d}y=\frac{12}{7}.$$

Similarly, take X^2 as the function of X and Y, then

$$EX^2=\iint\limits_{(x,y)\in G} x^2\,\frac{3xy}{16}\mathrm{d}x\mathrm{d}y=\int_0^2\mathrm{d}x\int_0^{x^2} x^2\,\frac{3xy}{16}\mathrm{d}y=3.$$

Finally, obtain that

$$Var(X)=E(X^2)-[E(X)]^2=3-\left(\frac{12}{7}\right)^2=\frac{3}{49}.$$

5.2.2 Expectations & Variance for Several Common Distributions

Theorem 5.3 **Expectation and variance of a Bernoulli random variable**

For a Bernoulli random variable $X\sim\text{Bernoulli}(p)$, we have

$$E(X)=p,$$
$$Var(X)=p(1-p).$$

Proof: The PMF of X is $p_X(x)=p^x(1-p)^{1-x}$ with $x=0,1;0<p<1$, thus

$$E(X)=1\cdot p+0\cdot(1-p)=p;$$
$$E(X^2)=1^2\cdot p+0\cdot(1-p)=p.$$

Finally, using the formula $Var(X)=E(X^2)-[E(X)]^2$, we obtain that

$$Var(X)=p-p^2=p(1-p).$$

Theorem 5.4 **Expectation and variance of a Binomial random variable**

For $X\sim\text{Bin}(n,p)$ we have

$$E(X)=np$$
$$Var(X)=np(1-p)$$

Proof: For Binomial random variable X with parameters n and p, it can be considered as the sum of n **independent** Bernoulli random variables with identical parameter p, i. e.

$$X=\sum_{i=1}^{n}X_i,X_i\sim\text{Bernoulli}(p).$$

For Bernoulli random variable, the expectation and variance are p and $p(1-p)$, hence,

$$EX=\sum_{i=1}^{n}E(X_i)=\sum_{i=1}^{n}p=np,$$

$$Var(X)=\sum_{i=1}^{n}Var(X_i)=\sum_{i=1}^{n}p(1-p)=np(1-p).$$

Theorem 5.5 **Expectation and variance of a Poisson random variable**

If $X\sim\text{Possion}(\lambda)$ then

$$E(X)=Var(X)=\lambda.$$

Proof: The PMF of X is

$$p_X(x)=\frac{e^{-\lambda}\lambda^x}{x!},x\in Z^+.$$

The mean of X can be calculated as follows:

$$EX=\sum_{x=0}^{\infty}xp_X(x)=\sum_{x=0}^{\infty}x\frac{e^{-\lambda}\lambda^x}{x!}=\lambda e^{-\lambda}\sum_{x=1}^{\infty}\frac{\lambda^{x-1}}{(x-1)!}=\lambda e^{-\lambda}e^{\lambda}=\lambda.$$

Similarly,

$$\begin{aligned}EX^2&=E[X(X-1)+X]=\sum_{x=0}^{\infty}x(x-1)p_X(x)+EX\\&=\sum_{x=0}^{\infty}x(x-1)\frac{e^{-\lambda}\lambda^x}{x!}+\lambda=\lambda^2e^{-\lambda}\sum_{x=2}^{\infty}\frac{\lambda^{x-2}}{(x-2)!}+\lambda\\&=\lambda^2+\lambda.\end{aligned}$$

Hence,

$$Var(X)=E(X^2)-(EX)^2=(\lambda^2+\lambda)-\lambda^2=\lambda.$$

Theorem 5.6 **Expectation and variance of a Uniform random variable**

If $X\sim U[a,b]$ then

$$E(X)=\frac{a+b}{2};$$

$$Var(X)=\frac{(b-a)^2}{12}.$$

Proof: The PDF of X is

$$f(x)=\begin{cases}1/(b-a), & a\leqslant x\leqslant b,\\0, & \text{otherwise.}\end{cases}$$

Using integration, we have

$$E(X)=\int_{-\infty}^{+\infty}xf_X(x)\mathrm{d}x=\int_a^b x\frac{1}{b-a}\mathrm{d}x=\frac{a+b}{2},$$

$$E(X^2)=\int_{-\infty}^{+\infty}x^2f_X(x)\mathrm{d}x=\int_a^b x^2\frac{1}{b-a}\mathrm{d}x=\frac{a^2+ab+b^2}{3}.$$

Hence,

$$Var(X)=E(X^2)-(EX)^2=\frac{a^2+ab+b^2}{3}-\left(\frac{a+b}{2}\right)^2=\frac{(b-a)^2}{12}.$$

Theorem 5.7 **Expectation and variance of a Exponential random variable**

If $X\sim Exp(\lambda)$ then

$$E(X)=1/\lambda,$$
$$Var(X)=1/\lambda^2.$$

Proof: The PDF of X is

$$f(x)=\begin{cases}\lambda e^{-\lambda x}, & x>0,\\0, & \text{otherwise.}\end{cases}$$

Using integration, we have

$$E(X)=\int_{-\infty}^{+\infty} x f_X(x)\mathrm{d}x=\int_{0}^{+\infty} x\lambda \mathrm{e}^{-\lambda x}\mathrm{d}x$$

$$=(-x\mathrm{e}^{-\lambda x})\Big|_0^{+\infty}+\int_0^{+\infty}\mathrm{e}^{-\lambda x}\mathrm{d}x=0-\frac{\mathrm{e}^{-\lambda x}}{\lambda}\Big|_0^{+\infty}=\frac{1}{\lambda},$$

$$E(X^2)=\int_{-\infty}^{+\infty} x^2 f_X(x)\mathrm{d}x=\int_{0}^{+\infty} x^2\lambda \mathrm{e}^{-\lambda x}\mathrm{d}x$$

$$=(-x^2\mathrm{e}^{-\lambda x})\Big|_0^{+\infty}+\int_0^{+\infty}2x\mathrm{e}^{-\lambda x}\mathrm{d}x=\frac{2}{\lambda^2}.$$

Hence,

$$Var(X)=E(X^2)-(EX)^2=\frac{2}{\lambda^2}-\frac{1}{\lambda^2}=\frac{1}{\lambda^2}.$$

Theorem 5.8 Expectation and variance of a Normal random variable

If $X\sim N(\mu,\sigma^2)$ then

$$E(X)=\mu,$$
$$Var(X)=\sigma^2.$$

Proof: The PDF of X is

$$f_X(x)=\frac{1}{\sqrt{2\pi}\sigma}\mathrm{e}^{-(x-\mu)^2/2\sigma^2},$$

which is symmetric around μ, so its expectation EX must be μ.

The variance is given by

$$Var(x)=\frac{1}{\sqrt{2\pi}\sigma}\int_{-\infty}^{+\infty}(x-\mu)^2\mathrm{e}^{-(x-\mu)^2/2\sigma^2}\mathrm{d}x$$

$$=\frac{1}{\sqrt{2\pi}\sigma}(-y\mathrm{e}^{-y^2/2\sigma^2})\Big|_{-\infty}^{+\infty}+\frac{\sigma^2}{\sqrt{2\pi}}\int_{-\infty}^{+\infty}\mathrm{e}^{-y^2/2}\mathrm{d}x,\ [\text{where } y=(x-\mu)/\sigma]$$

$$=0+\sigma^2$$

$$=\sigma^2.$$

【Example 5.10】 In Example 5.7, we suppose that X, Y and Z are independent random variables with expectations and variances are known. Here, if $X\sim$ Bin $(200,0.1)$, $Y\sim \mathrm{Exp}(0.1)$, $Z\sim N(2,4)$ and $W=X-2Y+5Z+1$. Try to obtain the expectation and variance of W.

According to Theorem 5.4, Theorem 5.7, Theorem 5.8, we have

$$EX=200\times 0.1=20, Var(X)=200\times 0.1\times 0.9=18,$$

$$EY=\frac{1}{0.1}=10, Var(Y)=\frac{1}{0.1^2}=100,$$

$$EZ=2, Var(Z)=4.$$

Hence,

$$
\begin{aligned}
EW &= EX-2EY+5EZ+1\\
&=20-2\times 10+5\times 2+1\\
&=11,\\
Var(W)&=Var(X)+(-2)^2 Var(Y)+5^2 Var(Z)+Var(1)\\
&=18+4\times 100+25\times 4+0\\
&=518.
\end{aligned}
$$

5.3 Covariance and Correlation Coefficient

5.3.1 Covariance and Correlation Coefficient

From the previous discussion of the joint PMF or joint PDF of two-dimensional random variable (X,Y), we know that when two random variables X and Y are not independent, there often exists relationship between X and Y. It is frequently of interest to assess how strongly they are related to one another. Is there any numerical feature to measure this relationship to some extent? Recalling contents discussed in Proposition 5.1, we have found that when X is independent of Y, there must have $E[(X-EX)(Y-EY)]=0$. That is, when $E[(X-EX)(Y-EY)]\neq 0$, X and Y are certainly not independent, so the numerical value $E[(X-EX)(Y-EY)]$ reflects some kind of relation between X and Y to some extent, we define it as covariance.

Definition 5.6 The **covariance**[1] between two random variables X and Y is

$$Cov(X,Y)=E[(X-EX)(Y-EY)]. \tag{5.11}$$

When $Cov(X,Y)=0$, we say that X and Y are uncorrelated.

That is, since $X-EX$ and $Y-EY$ are the deviations of the two variables from their respective mean values, the covariance is the expected product of deviations. From the definition of mixing moment of random variable, we find that the covariance is just the $1+1$ order mixing central moment.

Note: The variance of X is the covariance between X and itself, i.e.

$$Cov(X,X)=E[(X-EX)(X-EX)]=Var(X).$$

Proposition 5.4 (Shortcut formula for covariance)

$$Cov(X,Y)=E(XY)-EX\cdot EY. \tag{5.12}$$

The formula above has been proved in Proposition 5.2 (d), we omit it here.

From the definition of covariance, we can rewrite the formula of the variance

[1] covariance：协方差

for sum of X and Y.

Proposition 5.5 For any random variables X and Y, we have

$$Var(X+Y)=Var(X)+Var(Y)+Cov(X,Y)$$

According to the definition of covariance, it might appear that the relationship between X and Y is weak when the covariance value is small. But when X and Y increase k times at the same time, the relationship between kX and kY shouldn't be stronger than that between X and Y, which means that the covariance value won't change. However, it's not true.

Proposition 5.6 For any random variables X and Y, we have

$$Cov(kX,kY)=k^2Cov(X,Y).$$

Proof: Based on the shortcut formula of covariance, we have

$$\begin{aligned}Cov(kX,kY)&=E(k^2XY)-E(kX)\cdot E(kY)\\&=k^2[E(XY)-EX\cdot EY]\\&=k^2Cov(X,Y).\end{aligned}$$

This property shows a serious defect of covariance that makes it impossible to interpret a computed value. For example, when examining the relation between the price of almonds and the price of cashew nuts, the covariance is 0.3 if it is in "dollar" and 3000 if it is in "cent" because 1 dollar=100 cents. The defect of covariance is that its computed value depends critically on the units of measurement. Ideally, the choice of units should have no effect on a measure of strength of relationship. This is achieved by scaling the covariance.

Definition 5.7 The **correlation coefficient**[1] of X and Y is defined by

$$\rho_{X,Y}=\frac{Cov(X,Y)}{\sigma_X\cdot\sigma_Y}. \tag{5.13}$$

Which is also denoted by $Corr(X,Y)$ or just ρ.

It's clear to verify that the variation of the coefficients of the two random variables does not change the relationship between the two random variables.

Proposition 5.7 If a and b are either both positive or both negative, then

$$\rho_{X,Y}=\rho_{aX,bY}.$$

Proof: According to the definition of correlation coefficient, we have

$$\begin{aligned}\rho_{aX,bY}&=\frac{Cov(aX,bY)}{\sigma_{aX}\cdot\sigma_{bY}}\\&=\frac{E(aX\cdot bY)-E(aX)\cdot E(bY)}{\sqrt{Var(aX)}\sqrt{Var(bY)}}\end{aligned}$$

---

[1] correlation coefficient：相关系数

$$
\begin{aligned}
&= \frac{ab[E(XY)-EX \cdot EY]}{|a||b|\sqrt{Var(X)}\sqrt{Var(Y)}} \\
&= \frac{[E(XY)-EX \cdot EY]}{\sqrt{Var(X)}\sqrt{Var(Y)}} \\
&= \rho_{X,Y}.
\end{aligned}
$$

The penultimate equation holds because a and b are either both positive or both negative.

This proposition shows that ρ remedies the defect of $Cov(X,Y)$ and also suggests how to recognize the existence of a strong (linear) relationship.

【Example 5.11】 Let a two-dimensional random variable (X,Y) follows **bivariate normal distribution**, $N(\mu_1,\mu_2,\sigma_1^2,\sigma_2^2,\rho)$. Try to find the correlation coefficient $\rho_{X,Y}$.

Solution: As the joint PDF of (X,Y) is

$$
f(x,y)=\frac{1}{2\pi\sigma_1\sigma_2\sqrt{1-\rho^2}}e^{-\frac{1}{2(1-\rho^2)}\left[\frac{(x-\mu_1)^2}{\sigma_1^2}-2\rho\frac{(x-\mu_1)(y-\mu_2)}{\sigma_1\sigma_2}+\frac{(y-\mu_2)^2}{\sigma_2^2}\right]},
$$

$$
(-\infty<x<+\infty,-\infty<y<+\infty)
$$

and expectations of X and Y are μ_1 and μ_2, so

$$
\begin{aligned}
Cov(X,Y) &= E[(X-EX)(Y-EY)] \\
&= \frac{1}{2\pi\sigma_1\sigma_2\sqrt{1-\rho^2}}\int_{-\infty}^{+\infty}\int_{-\infty}^{+\infty}(x-\mu_1)(y-\mu_2)\times \\
&\quad \exp\left(\frac{-1}{2(1-\rho^2)}\left[\frac{(x-\mu_1)^2}{\sigma_1^2}-2\rho\frac{(x-\mu_1)(y-\mu_2)}{\sigma_1\sigma_2}+\frac{(y-\mu_2)^2}{\sigma_2^2}\right]\right)dy\,dx \\
&= \frac{1}{2\pi\sigma_1\sigma_2\sqrt{1-\rho^2}}\int_{-\infty}^{+\infty}\int_{-\infty}^{+\infty}(x-\mu_1)(y-\mu_2)e^{-\frac{(x-\mu_1)^2}{\sigma_1^2}} \cdot \\
&\quad e^{-\frac{1}{2(1-\rho^2)}\left(\frac{y-\mu_2}{\sigma_2}-\rho\frac{x-\mu_1}{\sigma_1}\right)^2}dy\,dx.
\end{aligned}
$$

Take a linear transformation: $\begin{cases} u=\dfrac{1}{\sqrt{1-\rho^2}}\left(\dfrac{y-\mu_2}{\sigma_2}-\rho\dfrac{x-\mu_1}{\sigma_1}\right) \\ v=\dfrac{x-\mu_1}{\sigma_1} \end{cases}$, where μ_1 and μ_2 are variances of X and Y respectively, we have

$$Cov(X,Y) = \frac{1}{2\pi}\int_{-\infty}^{+\infty}\int_{-\infty}^{+\infty}(\sigma_1\sigma_2\sqrt{1-\rho^2}\,uv + \rho\sigma_1\sigma_2 v^2)e^{-\frac{u^2+v^2}{2}}\,du\,dv$$

$$= \frac{\sigma_1\sigma_2\sqrt{1-\rho^2}}{2\pi}\left(\int_{-\infty}^{+\infty} u e^{-\frac{u^2}{2}}\,du\right)\left(\int_{-\infty}^{+\infty} v e^{-\frac{v^2}{2}}\,dv\right) +$$

$$\frac{\rho\sigma_1\sigma_2}{2\pi}\int_{-\infty}^{+\infty}\int_{-\infty}^{+\infty} v^2 e^{-\frac{u^2+v^2}{2}}\,du\,dv$$

$$= \frac{\rho\sigma_1\sigma_2}{2\pi}\left(\int_{-\infty}^{+\infty} e^{-\frac{u^2}{2}}\,du\right)\left(\int_{-\infty}^{+\infty} v^2 e^{-\frac{v^2}{2}}\,dv\right)$$

$$= \rho\sigma_1\sigma_2.$$

Hence,

$$\rho_{X,Y} = \frac{Cov(X,Y)}{\sigma_1 \cdot \sigma_2} = \frac{\rho\sigma_1\sigma_2}{\sigma_1 \cdot \sigma_2} = \rho.$$

Note:

- From above, we know that if (X,Y) follows bivariate normal distribution $(X,Y) \sim N(\mu_1, \mu_2, \sigma_1^2, \sigma_2^2, \rho)$, it follows (after some tricky integration) that the marginal distribution of X is normal with mean value μ_1 and variance σ_1^2, and similarly the marginal distribution of Y is normal with mean μ_2 and variance σ_2^2; The fifth parameter of the distribution is ρ, which can be shown to be the correlation coefficient between X and Y.
- For two normal random variables, uncorrelated means $\rho = 0$. On the other hand, $\rho = 0$ is equivalent to the independence of X and Y, which has been confirmed in Example 4.10. Therefore, as far as two random variables following **normal distribution** is concerned, **uncorrelated and independence are equivalent.**

5.3.2 The Essence of Covariance and Correlation Coefficient

Roughly speaking, a positive or negative covariance indicates that the values of $X-EX$ and $Y-EY$ obtained in a single experiment "tend" to have the same [see Figure 5.1(a)] or the opposite sign [see Figure 5.1(b)], respectively. Thus the sign of the covariance provides an important qualitative indicator of the relation between X and Y.

If X and Y are independent [as illustrated in Figure 5.2], then

$$Cov(X,Y) = E[(X-EX)(Y-EY)] = [E(X-EX)] \cdot [E(Y-EY)] = 0.$$

That is, if X and Y are independent, they are also uncorrelated. However, the reverse is not true, as illustrated by the following example.

【Example 5.12】 Suppose that the PMF of random variable X is as follows:

X	-1	0	1
p	1/3	1/3	1/3

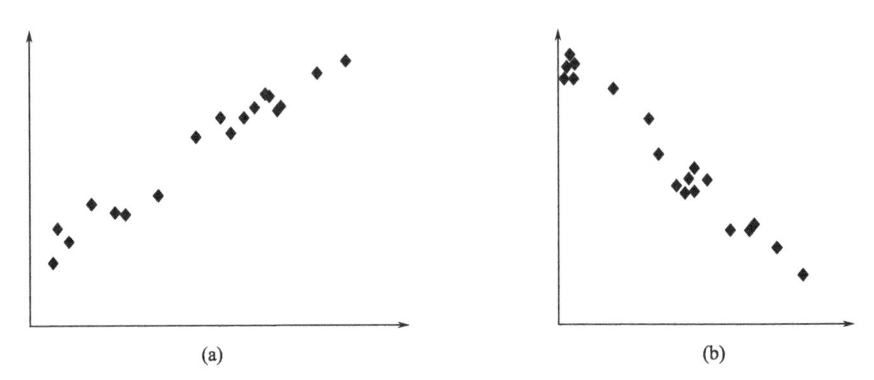

Figure 5.1 Examples of positively and negatively correlated random variables

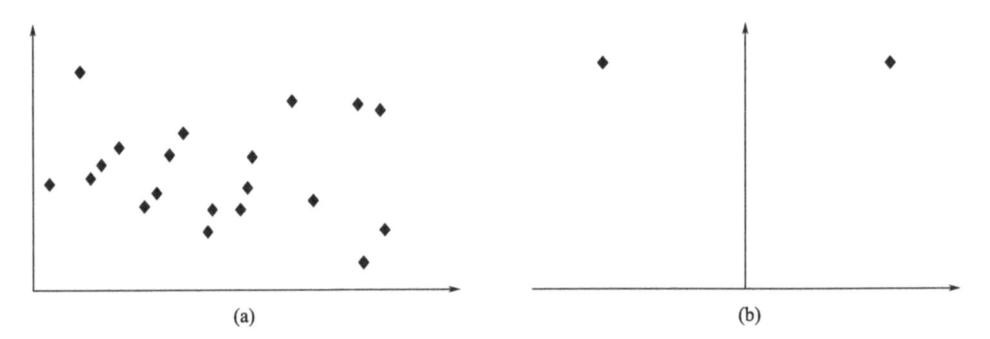

Figure 5.2 Examples of uncorrelated random variables

Another random variable Y is the function of X, $Y=X^2$. It's obvious that X and Y are **not independent** since a value of X fixes the value of Y.

Now, let's check the covariance of X and Y.

Since

$$EX=-1\times\frac{1}{3}+0\times\frac{1}{3}+1\times\frac{1}{3}=0;$$

$$E(XY)=E(X^3)=(-1)^3\times\frac{1}{3}+0^3\times\frac{1}{3}+1^3\times\frac{1}{3}=0.$$

Hence,

$$Cov(X,Y)=E(XY)-EX\cdot EY=0-0=0.$$

Thus,

$$\rho=\frac{Cov(X,Y)}{\sqrt{Var(X)}\sqrt{Var(Y)}}=0.$$

The result shows that X and Y are uncorrelated.

Remark:

- "Uncorrelated" and "independent" are not equal, though it's true for two normal random variables. In fact, **"Independent" is sufficient but not necessary for "uncorrelated".**

From above, we find that even closely related random variables (of course not independent) may have a zero correlation coefficient. Then, in what sense does the correlation coefficient represent the relationship between random variables? The following theorem will reveal the answer.

Theorem 5.9 For two random variables X and Y, if there exists correlation coefficient $\rho_{X,Y}$ between X and Y, then

(1) $|\rho_{X,Y}| \leqslant 1$;

(2) The necessary and sufficient condition for $|\rho_{X,Y}|=1$ is that X and Y are **linearly correlated**[1] by probability 1. That is, there are constants a, b, such that

$$P\{Y=aX+b\}=1;$$

where a and ρ are either both positive or both negative.

Proof: (1) Let $X_1=\dfrac{X-EX}{\sqrt{Var(X)}}, Y_1=\dfrac{Y-EY}{\sqrt{Var(Y)}}$, it's easy to verify that

$$\begin{cases} EX_1=EY_1=0; \\ Var(X_1)=EX_1^2=Var(Y_1)=EY_1^2=1; \\ E(X_1Y_1)=\rho_{X,Y}. \end{cases} \tag{5.14}$$

Consider a quadratic function of real variable t,

$$\begin{aligned} g(t) &= E(tX_1-Y_1)^2 \\ &= t^2 E{X_1}^2 - 2tE(X_1Y_1)+EY_1^2 \\ &= t^2-2\rho_{X,Y}t+1. \end{aligned}$$

The last equation can be obtained according to formula (5.14).

Since $g(t)$ is the expectation of the square of a random variable, it's certainly nonnegative, i.e. $g(t)\geqslant 0$, accordingly, the discriminant term Δ of quadratic equation $g(t)=0$ is non-positive, i.e.

$$\Delta=4{\rho_{X,Y}}^2-4\leqslant 0,$$

Hence,

$$|\rho_{X,Y}|\leqslant 1.$$

(2) If $|\rho_{X,Y}|=1$, we might as well assume that $\rho_{X,Y}=1$.

Then there is only one solution $t=1$ to the equation $g(t)=0$.

---

[1] linearly correlated：线性相关

In fact,

$$g(t)=E(tX_1-Y_1)^2=t^2-2t+1=(t-1)^2=0\Rightarrow t=1.$$

Hence,

$$g(1)=E(X_1-Y_1)^2=Var(X_1-Y_1)+[E(X_1-Y_1)]^2=Var(X_1-Y_1)+0=0.$$

From Proposition 5. 3-(5), $Var(X_1-Y_1)=0$ is equivalent to $P\{X_1-Y_1=c\}=1$ where c is a constant. And

$$\begin{aligned}P\{X_1-Y_1=c\}&=P\left\{\frac{X-EX}{\sqrt{Var(X)}}-\frac{Y-EY}{\sqrt{Var(Y)}}=c\right\}\\&=P\left\{Y=\frac{\sqrt{Var(Y)}}{\sqrt{Var(X)}}X-\frac{\sqrt{Var(Y)}}{\sqrt{Var(X)}}EX-c\sqrt{Var(Y)}+EY\right\}\\&=P\{Y=aX+b\}\\&=1.\end{aligned}$$

Where $a=\frac{\sqrt{Var(Y)}}{\sqrt{Var(X)}}>0, b=-\frac{\sqrt{Var(Y)}}{\sqrt{Var(X)}}EX-c\sqrt{Var(Y)}+EY.$

Similarly, when $\rho_{X,Y}=-1$, the equivalent result is that there exists constants a^*, b^*, where $a^*=-\frac{\sqrt{Var(Y)}}{\sqrt{Var(X)}}<0.$

Note:

- This proposition says that $\rho_{X,Y}$ is a measure of the degree of linear relationship between X and Y, and only when there is a perfect positive or negative linear relationship between the two variables will $\rho_{X,Y}$ be 1 or -1. However, if $|\rho_{X,Y}|\ll 1$, there may still be a strong relationship between the two variables, just one that is not linear (as is shown in Example 5. 11. And even if $|\rho_{X,Y}|$ is close to 1, it may be that the relationship is really nonlinear but can be well approximated by a straight line.

Exercises

1. The PMF of the amount of memory X (GB) in a purchased flash drive was given as

X	1	2	4	8	16
$p_X(x)$	0. 05	0. 10	0. 35	0. 40	0. 10

Compute the following numerical characteristics :

(1) $E(X)$;

(2) $Var(X)$ directly from the definition;

(3) The standard deviation of X;

(4) $Var(X)$ using the shortcut formula.

2. Let X be a Bernoulli random variable with parameter p,

   (1) Compute $E(X^2)$;

   (2) Show that $Var(X)=p(1-p)$;

   (3) Compute $E(X^{79})$.

3. Let Z be a random variable with probability density function

$$f_Z(z)=\begin{cases} -\frac{3}{2}z^2+\frac{3}{2}, & 0\leqslant z\leqslant 1, \\ 0, & \text{otherwise.} \end{cases}$$

   Find the expectation and standard deviation of Z.

4. A discrete random variable X has expectation 3 and variance 10. Let $Y=(X+1)^2$. Explain what is wrong with the following, and derive the correct expectation of Y.

$$\begin{aligned} EY=E(X+1)^2 &=E(X^2+2X+1) \\ &=E(X)^2+2E(X)+1 \\ &=3^2+2\times 3+1 \\ &=16. \end{aligned}$$

5. If $Y\sim\text{Poisson}(\lambda)$, prove that $E(Y^2)=\lambda^2+\lambda$. In your proof, you should not quote the result that $Var(Y)=\lambda$, but you may quote other results from the lecture notes.

6. For this question, you will need the result that a linear combination of a set of independent normal random variables is another normal random variable (so Z defined below has a normal distribution). Suppose $X_1\sim N(10,18)$ and $X_2\sim N(10,18)$, with X_1 and X_2 independent. Calculate an interval (a_1,b_1) such that $P\{Z\in(a_1,b_1)\}\approx 0.95$, with $Z=\frac{X_1+X_2}{2}$.

7. Let random variable X and Y are independent of each other and both follow Poisson distribution with parameter λ. Derive the probability mass function of $Z=X+Y$, and state the distribution of Z. You may quote the binomial theorem that

$$(1+1)^n=\sum_{a=0}^{n}\frac{n!}{a!(n-a)!}.$$

8. A cyclist leaves a bicycle chained to some railings, and returns five hours later to find that the bike has been stolen. Define T to be the time in which the bike

was stolen, counting in hours from when the bike was left by the owner. Assuming that T has a uniform distribution, calculate

(1) the mean of T and $E(T^2)$;

(2) the probability that T lies between 3 and 4 four hours;

(3) the probability that $T=2$.

9. (1) Let $Z \sim N(0,1)$. Find $E(e^Z)$.

(2) Following your result in part (1), if $\ln Y \sim N(\mu,\sigma^2)$, is it true that $E(Y)=e^{\mu}$?

10. Possible values of X, the number of components in a system submitted for repair that must be replaced, are 1, 2, 3, and 4 with corresponding probabilities 0.15, 0.35, 0.35, and 0 .15, respectively.

(1) Calculate $E(X)$ and then $E(5-X)$.

(2) Would the repair facility be better off charging a flat fee of \$75 or else the amount \$[150/(5−X)]? [Note: It is not generally true that $E(c/Y)=c/E(Y)$.

11. Suppose $E(X)=5$ and $E[X(X-1)]=27.5$. What is

(1) $E(X^2)$? {Hint: First verify that $E[X(X-1)]=E(X^2)-E(X)$}

(2) $Var(X)$?

(3) The general relationship among the quantities $E(X)$, $E[X(X-1)]$, and $Var(X)$?

12. Let X follows Poisson distribution with parameter 4. What is the probability that X exceeds its mean value by no more than one standard deviation?

13. An instructor has given a short quiz consisting of two parts. For a randomly selected student, let X = the number of points earned on the first part and Y=the number of points earned on the second part. Suppose that the joint PMF of X and Y is given in the accompanying table.

Y \ X	0	5	10	15
0	0.02	0.06	0.02	0.10
5	0.04	0.15	0.20	0.10
10	0.01	0.15	0.14	0.01

(1) If the score recorded in the grade book is the total number of points earned on the two parts, what is the expected recorded score $E(X+Y)$?

(2) If the maximum of the two scores is recorded, what is the expected recorded score?

14. Jack and Rose have agreed to meet for lunch between noon (0:00 p. m.) and

1:00 p.m. Denote Jack's arrival time by X, Rose's by Y, and suppose X and Y are independent with PDFs

$$f_X(x)=\begin{cases}3x^2, & 0\leqslant x\leqslant 1,\\ 0, & \text{otherwise},\end{cases} \quad f_Y(y)=\begin{cases}2y, & 0\leqslant y\leqslant 1,\\ 0, & \text{otherwise}.\end{cases}$$

15. Prove that $Cov(X,Y)=E(XY)-E(X)E(Y)$, and hence prove that $Cov(X, Y+Z)=Cov(X,Y)+Cov(X,Z)$.
16. Let X and Y be the heights (in cm) of two adult sisters.
    (1) Do you think X and Y should be independent?
    (2) Do you think $X-Y$ and $X+Y$ should be independent?
    (3) If $Var(X)=Var(Y)$, find $Cov(X+Y,X-Y)$. Comment on your result.
17. Three people are living in one house, and in Week 1, are all at risk of catching a cold. Suppose that each person has a probability of 0.1 of catching a cold in Week 1, independently of the others in the house. If no-one catches a cold, the probabilities of catching a cold are unchanged for Week 2. If at least one person catches a cold in Week 1, the remaining 'uninfected' housemates each have a probability of 0.2 of catching a cold in Week 2, independently of the others in the house.
    Assume that no-one will catch a cold twice (e.g if all three housemates catch a cold in Week 1, no-one can catch a cold in Week 2).
    (1) Tabulate the joint probability mass function of the number of housemates who catch a cold in each of Weeks 1 and 2.
    (2) Calculate the correlation between the numbers of housemates who catch a cold in each of Weeks 1 and 2.
18. (1) Use the rules of expected value to show that $Cov(aX+b,cY+d)=ac(X,Y)$.
    (2) Use part (1) along with the rules of variance and standard deviation to show that $Corr(aX+b,cY+d)=Corr(X,Y)$ when a and c have the same sign.
    (3) What happens if a and c have opposite signs?
19. Suppose that the joint PMF of X and Y is given in the accompanying table.

Y \ X	0	5	10	15
0	0.02	0.06	0.02	0.10
5	0.04	0.15	0.20	0.10
10	0.01	0.15	0.14	0.01

Compute correlation coefficient ρ for X and Y.

20. Suppose that the joint PDF of X and Y is given as follows:

$$f_{X,Y}(x,y)=\begin{cases}8xy, & 0\leqslant x\leqslant y\leqslant 1,\\ 0, & \text{otherwise.}\end{cases}$$

Compute the correlation coefficient ρ for this X and Y.

21. Show that if $Y=aX+b(a\neq 0)$, then $\rho_{X,Y}=1$ or -1. Under what conditions will $\rho_{X,Y}=1$?

Chapter 6
Sums of Random Variables

6.1 Sums of Independent and Identically Distributed Random Variables

Suppose a sequence of random variables $X_1, X_2, \cdots, X_n$, each having the same probability distribution, with all the random variables independent of each other. Recall that by independent, which mean

$$P\{\{X_i \leqslant x_i\} \cap \{X_j \leqslant x_j\}\} = P\{X_i \leqslant x_i\} P\{X_j \leqslant x_j\}, \tag{6.1}$$

or equivalently,

$$P\{X_i \leqslant x_i | X_j \leqslant x_j\} = P\{X_i \leqslant x_i\}, \tag{6.2}$$

for $i \neq j$ in both cases. We say that $X_1, X_2, \cdots, X_n$ are independent and identical distribution (i. i. d). Now let's define $S(n)$ to be the sum and $\overline{X}(n)$ to be the mean of the samples.

$$S(n) = \sum_{i=1}^{n} X_i, \tag{6.3}$$

And

$$\overline{X}(n) = \frac{S(n)}{n}. \tag{6.4}$$

Both $S(n)$ and $\overline{X}(n)$ are also random variables, as they are functions of the random variables $X_1, \cdots, X_n$. If we write $E(X_i) = \mu$ and $Var(X_i) = \sigma^2$ (for all i, as the variables are i. i. d.), it is straightforward to derive the mean and variance of $S(n)$ and $\overline{X}(n)$ in terms of μ and σ^2.

Firstly,

$$E\{S(n)\} = \sum_{i=1}^{n} EX_i = n\mu. \tag{6.5}$$

Also, noting that $Cov(X_i, X_j)=0$ for $i \neq j$,

$$Var\{S(n)\} = \sum_{i=1}^{n} DX_i = n\sigma^2. \tag{6.6}$$

Then we have

$$E\{\overline{X}(n)\} = \frac{\sum_{i=1}^{n} EX_i}{n} = \mu, \tag{6.7}$$

and

$$Var\{\overline{X}(n)\} = \frac{\sum_{i=1}^{n} DX_i}{n^2} = \frac{\sigma^2}{n}. \tag{6.8}$$

These two results have important applications in Statistics. Suppose we are able to observe i. i. d. variables $X_1, \cdots, X_n$, but we don't know the value of $E(X_i)=\mu$. Equation (6.8) tells us that as n increases, the variance of $\overline{X}(n)$ gets smaller, and the smaller the variance is, the closer we expect $\overline{X}(n)$ to be to its mean value. Equation (6.7) tells us that the mean value of $\overline{X}(n)$ is μ (for any value of n). In other words, as n gets larger, we expect $\overline{X}(n)$ to be increasingly close to the unknown quantity μ, so we can use the observed value of $\overline{X}(n)$ to *estimate* μ.

We illustrate it in Figure 6.1. The four plots show the density functions of

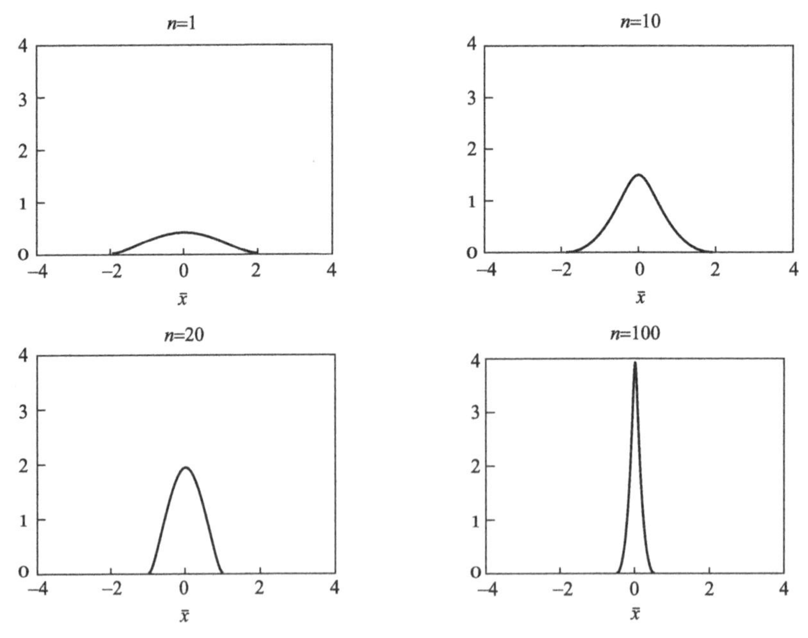

Figure 6.1 The density function of $\overline{X}(n)$, when $X_i \sim N(0,1)$, for four choices of n.

$\overline{X}(n)$ for $n=1,10,20$ and 100. In each case, $X_1,\cdots,X_n \sim N(0,1)$, so $E(X_i)=\mu=0$. We can see that the density function of $\overline{X}(n)$ becomes more tightly concentrated about the value 0 as n increases, and that the observed value of $\overline{X}(n)$ is increasingly likely to be close to 0.

If we didn't know that $\mu=E(X_i)=0$, but we could observe $\overline{X}(n)$, then using the observed value of $\overline{X}(n)$ for large n is likely to give us a good estimate of μ, as $\overline{X}(n)$ is likely to be close to $\mu(\mu=0)$.

6.2 Laws of Large Numbers[1]

We will now derive another important result regarding the behavior of $\overline{X}(n)$ for a large n. We firstly prove a useful inequality.

6.2.1 Chebyshev's Inequality[2]

Theorem 6.1 Let X be a random variable, with range R, $E(X)=\mu$ and $Var(X)=\sigma^2$. Then for any $c \geqslant 0$

$$P(|X-\mu| \geqslant c) \leqslant \frac{\sigma^2}{c^2}. \tag{6.9}$$

This inequality is called the Chebyshev's inequality.

Proof: Here we just take the continuous random variable as example. The density of the variable is $f(x)$.

So $$P\{|X-\mu| \geqslant c\} = \int\limits_{|x-\mu| \geqslant c} f(x)\mathrm{d}x \leqslant \int\limits_{|x-\mu| \geqslant c} \frac{|x-\mu|^2}{c^2} f(x)\mathrm{d}x$$

$$\leqslant \frac{1}{c^2}\int_{-\infty}^{+\infty} (x-\mu)^2 f(x)\mathrm{d}x = \frac{\sigma^2}{c^2}, \tag{6.10}$$

and the result follows.

6.2.2 The Weak Law of Large Numbers[3]

Let $X_1, X_2, \cdots$ be a sequence of i. i. d. random variables, each with mean μ and variance σ^2. Then for all $\varepsilon > 0$,

$$P(|\overline{X}(n)-\mu| \geqslant \varepsilon) \leqslant \frac{\sigma^2}{\varepsilon^2 n}. \tag{6.11}$$

---

[1] laws of large numbers：大数定律

[2] Chebyshev's inequality：切比雪夫不等式

[3] the weak law of large numbers：弱大数定律

This follows from Chebyshev's inequality, and the fact that

$$E\{\overline{X}(n)\}=\mu \text{ and } Var\{\overline{X}(n)\}=\frac{\sigma^2}{n}.$$

From equation (6.11), we have

$$\lim_{n\to\infty} P(|\overline{X}(n)-\mu|\geqslant\varepsilon)=0. \tag{6.12}$$

We can choose ε to be as small as we would like, so as n increases, it becomes increasingly unlikely that $\overline{X}(n)$ will differ from μ by any amount ε. Equation (6.12) is known as the **weak law of large numbers.**

It is possible to prove a stronger result, which is

$$P(\lim_{n\to\infty}\overline{X}(n)=\mu)=1, \tag{6.13}$$

This result is known as **the strong law of large numbers**[1].

6.3 The Central Limit Theorem (CLT)[2]

We finish with another important result which tells about the distribution of $\overline{X}(n)$ for large n.

Theorem 6.2 **(The central limit theorem)**

Let $X_1, X_2, \cdots$ be a sequence of i. i. d random variables, each with mean μ and variance σ^2. For any $-\infty\leqslant a\leqslant b<+\infty$,

$$\lim_{n\to\infty} P\left(a\leqslant\frac{\overline{X}(n)-\mu}{\sigma/\sqrt{n}}\leqslant b\right)=\frac{1}{\sqrt{2\pi}}\int_a^b \exp\left(-\frac{1}{2}z^2\right)\mathrm{d}z. \tag{6.14}$$

In other words, the distribution of $\overline{X}(n)$ tends to a normal distribution with mean μ and variance σ^2/n, as $n\to\infty$. Since $S(n)=n\times\overline{X}(n)$, for large n, we have, approximately

$$\overline{X}(n)\sim N\left(\mu,\frac{\sigma^2}{n}\right), \tag{6.15}$$

$$S(n)\sim N(n\mu, n\sigma^2). \tag{6.16}$$

As long as $X_1, X_2, \cdots$ have the same distribution, it doesn't matter what that distribution is. For example, even $X_1, X_2, \cdots$ are discrete random variables, the distribution of $\overline{X}(n)$ will still tend to a normal distribution and $\frac{\overline{X}(n)-\mu}{\sigma/\sqrt{n}}$ will tend

[1] the strong law of large numbers：强大数定律

[2] the central limit theorem (CLT)：中心极限定理

to a standard normal distribution.

We have already established that $E\{\overline{X}(n)\}=\mu$ and $Var\{\overline{X}(n)\}=\sigma^2/n$ (for any value of n), so the CLT is giving us the extra result that the distribution of $\overline{X}(n)$ tends to a normal distribution. A proof of the CLT is outside the scope of this course.

6.3.1 Example: Sums of Exponential Random Variables

Let $X_1, X_2, \cdots$ be a sequence of i. i. d exponential random variables, each with rate parameter 1. Note that the shape of the Exp(rate=1) distribution is very 'non-normal' (compare the pdf of an exponential random variable with that of a normal random variable). Now $\mu=E(X_i)=1$ and $\sigma^2=Var(X_i)=1$. For a large n, approximately, we should have

$$\overline{X}(n) \sim N\left(1, \frac{1}{n}\right).$$

Now we do a simulation experiment using R to compare this approximation with what we get if we simulate random values of $\overline{X}(n)$ lots of times, for different values of n. The results are shown in Figure 6.2. In the top left plot, we have the case $n=1$. The solid line shows the $N(1,1)$ density function, and the histogram represents the distribution of $\overline{X}(1)$ (which is just a single value of X), based on what we see when we simulate random values of $\overline{X}(1)$ many times. The histogram doesn't match the shape of the density function, which is to be expected, as we know that the density function of a single exponential random variable looks nothing like a 'bell-shaped' curve.

In the top right plot, we repeat the experiment, but now with $n=10$. Now we can see that histogram of simulated values of $\overline{X}(10)$ is closer in shape to the $N(1,1/10)$ density function, so even with small n, the CLT is giving a good approximation for the distribution of $\overline{X}(n)$. Larger values of n are tested in the remaining two plots.

In Figure 6.2, the solid line shows the density function under the CLT approximation. The histogram represents the distributed for simulated values of $\overline{X}(n)$, from a simulation in R.

6.3.2 Example: Sums of Bernoulli Random Variables, and the Normal Approximation to the Binomial Distribution

Let $X_1, X_2, \cdots$ be a sequence of i. i. d Bernoulli random variables, with $P(X_i=1)=p$ for each i. Then $\mu=E(X_i)=p$, and $\sigma^2=Var(X_i)=p(1-p)$.

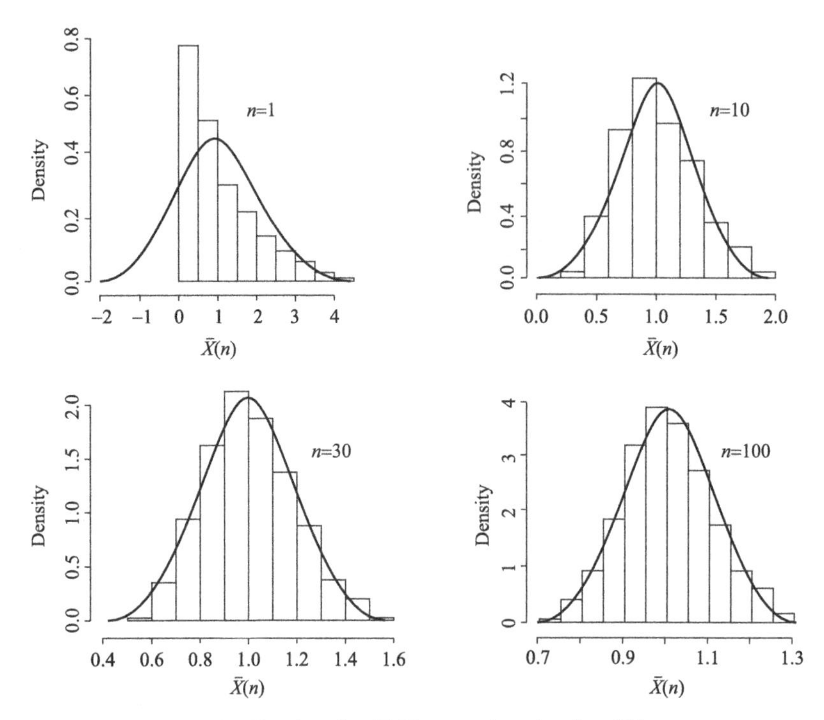

Figure 6.2 Testing the CLT approximation for different n

Then by the Central Limit Theorem, $\overline{X}(n)$ tends to a normal distribution with mean p and variance $p(1-p)/n$.

By definition $S(n)$, the total number of 'successes', has a Binomial distribution, so as $n \to \infty$, the distribution of $S(n)$ tends to a normal distribution, with

$$E\{S(n)\}=np$$

and

$$Var\{S(n)\}=np(1-p).$$

So that for large n, a Binomial random variable is approximately normally distributed, with mean np and variance $np(1-p)$.

【Example 6.1】 A man claims to have psychic abilities, in that he can predict the outcome of a coin toss. If he is tested 100 times, but he is really just guessing, what is probability he will be right 60 or more times?

Let X be the number of correct predictions. If the man is guessing, and the coin is fair, the probability that he is right on any single occasion is 0.5. Then

$$X \sim \text{Binomial}(100, 0.5).$$

Approximately, using the CLT,

$$X \sim N(50, 25)$$

so

$$P(X\geqslant 60)=1-\Phi\left(\frac{60-50}{\sqrt{25}}\right)=1-\Phi\left(\frac{10}{5}\right)=1-\Phi(2)=0.0228.$$

Exercises

1. 50 patients with gastroesophageal reflux disease are treated with a new drug to relieve pain from heartburn. Following treatment, the time T_i until patient i next experiences the symptoms is recorded. The doctor treating the patients thinks there is a 50% chance that patient i will stay symptom-free for at least 30 days. Assuming the times are independent and identically distributed, each with an exponential distribution, find the expectation and variance of

$$\overline{T}(50)=\frac{1}{50}\sum_{i=1}^{50}T_i.$$

2. Prove Markov's inequality, that for a continuous random variable X with range $R_X=[0,\infty]$ and for any $c>0$,

$$P(X>c)\leqslant\frac{E(X)}{c}.$$

3. Suppose, for a lottery scratch card, probabilities of different prizes are as follows:

prize	probability
£0	0.689
£2	0.300
£10	0.010
£100	0.001

If a scratch card costs £1, and you buy 100 scratch cards, calculate (approximately) the probability that you would make a profit.

4. Let X_k $(k\geqslant 1)$ is independent, with $E(X_k)=\mu$, $\sqrt{D(X_k)}=2.4$, and $S_n=\sum_{k=1}^{n}X_k$. Use the Central Limit Theorem to determine how large must n be to satisfy $P\left(\left|\frac{S_n}{n}-\mu\right|\leqslant 0.2\right)\geqslant 0.98$.

5. When the calculator performs an addition, it rounds each addition number to the integer closest to it. Let all the rounding errors be independent and uniformly distributed on $(-0.5,0.5)$. Now that 1500 numbers are added, what is

the probability that the absolute value of the sum of errors exceeds 15?

6. There are 200 families in one apartment. The number of distribution of each family's cars is as follows:

X	0	1	2
p_k	0.1	0.6	0.3

How many parking spaces is required to make the probability of each car has a parking space equals to 0.95 at least?

Chapter 7 Random Samples and Sampling Distributions

7.1 Random Sampling[1]

Sampling is one of the most important concepts in the study of statistics. In this chapter, we focus on sampling from distributions or populations and study such important quantities as the sample mean and sample variance, which will be of vital importance in future chapters.

The totality of observations with which we are concerned, whether their number be finite or infinite, constitutes what we call a population. There was a time when the world population referred to observations obtained from statistical studies about people. Today, statisticians use the term to refer to observations relevant to anything of interest.

Definition 7.1 A **population**[2] consists of the totality of the observations with which we are concerned.

The number of observations in the population is defined to be **the size of the population.** The numbers on the cards in a deck, the blood type of all students in a school are examples of populations with finite size. The observations obtained by measuring the atmospheric pressure every day, from the past on into the future is the example of population whose size is infinite. Some finite populations are so large that in theory we assume them to be infinite. For example, the population of

---

[1] Random Sampling：随机抽样

[2] population：总体

lifetimes of a certain type of storage battery being manufactured for mass distribution throughout the country.

Each observation in a population is a value of a random variable X having some probability distribution $f(x)$, we call random variable X **the population random variable X.** In the blood type experiment, the random variable X represents the type of blood and is assumed to take on values from 1 to 8. The lives of the storage batteries are values assumed by a continuous random variable having perhaps a normal population. Hereafter, when we refer to a "binomial population", a "normal population", or the "population $f(x)$", we shall mean a population whose observations are values of a random variable having a binomial distribution, a normal distribution, or the distribution $f(x)$.

It is impossible or impractical to observe the entire set of observations that make up the population. Therefore, we must depend on a subset of observations from the population to help us make inference concerning that population. This brings us to consider the notion of sampling.

Definition 7.2 A **sample**❶ is a subset of a population. The process of performing an experiment to obtain a sample from the population is called **sampling.**

The purpose of the sampling is to find out something about the nature of the population. If our inferences from the sample to the population are to be valid, we must obtain samples that are representative of the population. We are often tempted to choose a sample by selecting the most convenient members of the population. Such a procedure may lead to erroneous inferences concerning the population. To eliminate the possibility of erroneous inferences, it is desirable to choose a random sample in the sense that the observations are made independently and at random.

Definition 7.3 Let $X_1, X_2, \cdots, X_n$ be n independent random variables, each having the same probability distribution $f(x)$. Define $X_1, X_2, \cdots, X_n$ to be a **random sample**❷ of size n from the population $f(x)$.

In selecting a random sample of size n from a population $f(x)$, let us define the random variable $X_i, i=1, 2, \cdots, n$, to represent the ith measurement of sample value that we observe. The random Variables $X_1, X_2, \cdots, X_n$ will then constitute a random sample from the population $f(x)$ with numerical values $x_1, x_2, \cdots, x_n$ if the measurements are obtained by repeating the experiment n independent times under essentially the same conditions. Because of the identical conditions under

---

❶ sample：样本

❷ random sample：随机样本

which the elements of the sample are selected, it is reasonable to assume that the n random variables $X_1, X_2, \cdots, X_n$ are independent and that each has the same probability distribution $f(x)$.

For example, we make a random selection of 9 storage batteries from a manufacturing process that has maintained the same specification throughout and records the length of life for each battery, With the first measurement x_1 being a value of X_1, the second measurement x_2 being a value of X_2, and so forth, then x_1, $x_2, \cdots, x_8$ are the values of the random sample $X_1, X_2, \cdots, X_8$. If we assume the population of battery is normal, the values of any $X_i, i=1,2,\cdots,8$ will be precisely the same as those in the original population, and hence X_i has the same normal distribution as population random variable X.

If the population random variable X has the probability distribution $f(x)$, then random sample $X_1, X_2, \cdots, X_n$ from the population $f(x)$ are independent and each has the same probability distribution $f(x)$. That is, the probability distributions of $X_1, X_2, \cdots, X_n$ are $f(x_1), f(x_2), \cdots, f(x_n)$, and their joint probability distribution is

$$f(x_1, x_2, \cdots, x_n) = f(x_1)f(x_2)\cdots f(x_n). \tag{7.1}$$

7.2 Some Important Statistics

Our main purpose in selecting random samples is to elicit information about the unknown population parameters. For example, we wish to arrive at a conclusion concerning the proportion of tea-drinkers in China who prefer a certain brand of tea. It would be impossible to question every tea-drinkers which brand they prefer. Instead, a large random sample is selected and the proportion $\hat{p}$ of people in this sample favoring the brand of tea in question is calculate. The value $\hat{p}$ is used to make an inference concerning the true proportion p. $\hat{p}$ is a function of the observed values in the random sample. Since many random samples are possible from the same population, we would expect $\hat{p}$ to vary somewhat from sample to sample. That is, $\hat{p}$ is a value of a random variable that we represent by P. Such a random variable is called a statistic.

Definition 7.4 Any function of the random variables constituting a random sample which does not depend on any unknown parameters is called a **statistic**[1].

---

[1] statistic：统计量

Statistic is a random variable and hence has its own probability distribution, moments and other characteristics. It is a quantity whose observed value can be calculated once the sample has been chosen. We will introduce some basic statistics in this section.

7.2.1 Location Measures of a Sample

The most commonly used statistics for measuring the center of a set of data are the mean, median and mode.

Let $X_1, X_2, \cdots, X_n$ be a random sample of size n from a population random variable X, $x_1, x_2, \cdots, x_n$ are the values of observations of the random sample $X_1, X_2, \cdots, X_n$, then

(1) The **sample mean**[1] is

$$\overline{X}=\frac{1}{n}\sum_{i=1}^{n}X_i. \tag{7.2}$$

The observation value of $\overline{X}$ is denoted by $\overline{x}=\frac{1}{n}\sum_{i=1}^{n}x_i$. The term sample mean is applied to both the statistic $\overline{X}$ and its observation value $\overline{x}$. In the future, normally we use capital letters to represent random variables and use small letters to represent the observed values.

(2) The **sample median**[2]:

Ranking the elements of the random sample $X_1, X_2, \cdots, X_n$ in an increasing order to yield $X_{(1)}, X_{(2)}, \cdots, X_{(n)}$, where $X_{(1)}$ is the smallest, $X_{(n)}$ is the largest, then the sample median is

$$\hat{M}_d=\begin{cases} X_{\left(\frac{n+1}{2}\right)}, & \text{if } n \text{ is odd,} \\ \frac{1}{2}\left[X_{\left(\frac{n}{2}\right)}+X_{\left(\frac{n}{2}+1\right)}\right], & \text{if } n \text{ is even.} \end{cases} \tag{7.3}$$

The sample median is also a location measure that shows the middle value of the sample.

(3) The **sample mode**[3] is the value of the sample that occurs most often. If the highest frequency is shared by multiple values, then there are multiple modes.

【Example 7.1】 Suppose a random sample $X_1, X_2, \cdots, X_{10}$ has the following observation values: 0.32, 0.53, 0.28, 0.37, 0.47, 0.43, 0.36, 0.42, 0.38, 0.43. Find the sample mean, sample median and sample mode.

---

[1] sample mean：样本均值

[2] sample median：样本中位数

[3] sample mode：样本众数

Solution: We can compute the following.

Sample mean: $\bar{x}=\frac{1}{10}(0.32+0.53+0.28+0.37+0.47+0.43+0.36+0.42+0.38+0.43)=0.399$.

The observation values of $X_{(1)}, X_{(2)}, \cdots, X_{(n)}$ are 0.28, 0.32, 0.36, 0.37, 0.38, 0.42, 0.43, 0.43, 0.47. 0.53.

Sample median: $\hat{m}_d=\frac{1}{2}(x_{(5)}+x_{(6)})=\frac{1}{2}(0.38+0.42)=0.4$.

Sample mode: 0.43.

7.2.2 Variability Measures of a Sample

The variability in a sample displays how the observations spread out from the average. It is possible to have two sets of observations with the same mean or median that differ in the variability of their measurements about the average.

Consider the following measurements, in liters, for two samples of orange juice bottled by companies A and B:

Sample A	0.97	1.00	0.94	1.03	1.06
Sample B	1.06	1.01	0.88	0.91	1.14

Both samples have the same mean, 1.00 liter. But it is obvious that company A bottles orange juice with a more uniform content than company B. We say that the variability of the observations from the average is less for sample A than for sample B. Therefore, in buying orange juice, we feel more confident that the bottles we select will be close to the advertised average if we buy from company A.

We shall introduce several measures of sampling variability.

(1) The **sample variance**[1]

$$S^2=\frac{1}{n-1}\sum_{i=1}^{n}(X_i-\bar{X})^2. \tag{7.4}$$

The computed value of S^2 for a given sample is denoted by s^2, that is $s^2=\frac{1}{n-1}\sum_{i=1}^{n}(x_i-\bar{x})^2$. The reason for using $n-1$ as divisor rather than the more obvious choice n will become apparent in Chapter 8.

(2) The **sample standard deviation**[2]:

$$S=\sqrt{S^2}\text{, where } S^2 \text{ is the sample variance.} \tag{7.5}$$

---

[1] sample variance：样本方差

[2] sample standard deviation：样本标准误差

(3) The **sample range**[1]:

$$R = X_{(n)} - X_{(1)}. \tag{7.6}$$

【Example 7.2】 Find the variance, standard deviation and range of the data 3, 4, 5, 6, 6 and 7, representing the number of trout caught by a random sample of 6 fishermen at Lake Muskoka.

Solution: Calculating the sample mean, we get

$$\bar{x} = \frac{1}{6}(3+4+5+6+6+7) = \frac{31}{6}.$$

Therefore,

$$s^2 = \frac{1}{5}\sum_{i=1}^{6}\left(x_i - \frac{31}{6}\right)^2 = \frac{1}{5}\left[\left(3-\frac{31}{6}\right)^2 + \left(4-\frac{31}{6}\right)^2 + \left(5-\frac{31}{6}\right)^2 + \left(6-\frac{31}{6}\right)^2 + \left(6-\frac{31}{6}\right)^2 + \left(7-\frac{31}{6}\right)^2\right]$$

$$= \frac{1}{5}\left(\frac{169}{36} + \frac{49}{36} + \frac{1}{36} + \frac{25}{36} + \frac{25}{36} + \frac{121}{36}\right) = \frac{1}{5} \cdot \frac{390}{36}$$

$$= \frac{13}{6}.$$

Thus, the sample standard deviation $s = \sqrt{13/6} = 1.47$ and the sample range $r = 7-3=4$.

Rather than computing the sample variance directly from the definition, we can use a computational formula which involves less numerical effort.

【Example 7.3】 If S^2 is the variance of a random sample of size n, we may write

$$S^2 = \frac{1}{n-1}\left(\sum_{i=1}^{n} X_i^2 - n\bar{X}^2\right) \tag{7.7}$$

Proof: Note that

$$\sum_{i=1}^{n}(X_i - \bar{X})^2 = \sum_{i=1}^{n}(X_i^2 - 2X_i\bar{X} + \bar{X}^2) = \sum_{i=1}^{n} X_i^2 - 2\bar{X}\sum_{i=1}^{n} X_i + n\bar{X}^2$$

Since $\sum_{i=1}^{n} X_i = n\bar{X}$, We have

$$\sum_{i=1}^{n}(X_i - \bar{X})^2 = \sum_{i=1}^{n} X_i^2 - n\bar{X}^2$$

From which the conclusion follows directly.

For example, we can find the variance of the data in Example 7.2 as follows.

We find that $\sum_{i=1}^{6} x_i^2 = 171, \bar{x} = \frac{31}{6}$, and $n=6$. Hence,

---

[1] sample range: 样本极差

$$s^2=\frac{1}{6-1}\left(\sum_{i=1}^{6}x_i^2-6\overline{x}^2\right)=\frac{1}{5}\left[171-6\times\left(\frac{31}{6}\right)^2\right]=\frac{13}{6}.$$

7.3 Sampling Distributions

A statistic varies or changes for each different random sample; That is, it is a random variable. The probability distribution of the statistic is called **sampling distribution**[1].

Definition 7.5 The **sampling distribution of a statistic** is the probability distribution for the possible values of the statistic that results when random samples of size n are repeatedly drawn from the population.

In this chapter we study several of the important sampling distributions of frequently used statistics. Applications of these sampling distributions to problems of statistical inference are considered throughout the following chapters. The first important sampling distribution to be considered is that of the mean $\overline{X}$. The probability distribution of $\overline{X}$ is called the sampling distribution of the mean.

Theorem 7.1 If $\overline{X}$ is the mean of the random sample $X_1, X_2, \cdots, X_n$ of size n from a random variable X which has mean μ and the variance σ^2, then

$$E(\overline{X})=\mu \text{ and } D(\overline{X})=\frac{\sigma^2}{n}. \tag{7.8}$$

Proof: First,

$$E(\overline{X})=E\left(\frac{1}{n}\sum_{i=1}^{n}X_i\right)=\frac{1}{n}E\left(\sum_{i=1}^{n}X_i\right)=\frac{1}{n}\sum_{i=1}^{n}\mu=\mu.$$

Second, since $X_1, X_2, \cdots, X_n$ are independent, we can get that

$$D(\overline{X})=D\left(\frac{1}{n}\sum_{i=1}^{n}X_i\right)=\frac{1}{n^2}\sum_{i=1}^{n}D(X_i)=\frac{1}{n^2}\sum_{i=1}^{n}\sigma^2=\sigma^2.$$

It is customary to write $E(\overline{X})$ as $\mu_{\overline{X}}$ and $D(\overline{X})$ as $\sigma_{\overline{X}}^2$. Here, $\sigma_{\overline{X}}=\frac{\sigma}{\sqrt{n}}$ is called the standard error of the mean, which shows that the standard deviation of the distribution of $\overline{X}$ decrease when n is increased. It means that when n becomes larger, we can expect that the value of $\overline{X}$ to be closer to μ.

We are often interested in sampling from a population that follows a normal

[1] sampling distribution：抽样分布

distribution. The following result, stated without a proof, shows that the distribution of $\overline{X}$ in the case.

Corollary 7.1 If $\overline{X}$ is the mean of a random sample of size n from a normal population with mean μ and variance σ^2, then $\overline{X} \sim N\left(\mu, \frac{\sigma^2}{n}\right)$.

【Example 7.4】 An electrical firm manufactures light bulbs that have a length of life that is normally distributed, with mean equal to 800 hours and a standard deviation of 40 hours. Find the probability that a random sample of 16 bulbs will have an average life of less than 775 hours.

Solution: In this case, $\mu=800$, $\sigma=40$ and $n=16$.

Since the average life $\overline{X} \sim N\left(800, \frac{40^2}{16}\right)$, so we have

$$P\{\overline{X}<775\}=P\left\{\frac{\overline{X}-800}{40/\sqrt{16}}<\frac{775-800}{40/\sqrt{16}}\right\}=\Phi(-2.5)=1-\Phi(2.5)$$
$$=1-0.9938=0.0062.$$

7.4 Some Important Sampling Distribution

When the distribution function of the population is known, the sampling distribution is determined. However, it is generally difficult to obtain the exact distribution of statistics. This section describes the distribution of several common statistics from normal populations.

7.4.1 Chi-square Distribution[1]

There is a probability density, **gamma distribution**[2], which plays an important role in statistics.

Definition 7.6 A random variable X has a gamma distribution if its probability density function is given by

$$f(x)=\begin{cases}\dfrac{1}{\beta^{\alpha}\Gamma(\alpha)}x^{\alpha-1}e^{-\frac{x}{\beta}}, & \text{for } x>0,\\ 0, & \text{for } x\leqslant 0,\end{cases} \tag{7.9}$$

where $\alpha>0, \beta>0$.

In the future, we will write that $X \sim \Gamma(\alpha, \beta)$ if a random variable X has a

[1] Chi-square distribution: 卡方分布

[2] gamma distribution: 伽马分布

gamma distribution with the parameters α and β.

Gamma distribution has some properties that will be useful in later discussions. We list them in the following theorems without proofs.

Theorem 7.2 The mean and variance of the gamma distribution $X \sim \Gamma(\alpha, \beta)$ are given by

$$E(X)=\alpha\beta \text{ and } D(X)=\alpha\beta^2. \tag{7.10}$$

Theorem 7.3 If $X_1, X_2, \cdots, X_n$ are independent random variables and $X_i \sim \Gamma(\alpha_i, \beta), i=1,2,\cdots,n$, then

$$X_1+X_2+\cdots+X_n \sim \Gamma(\alpha_1+\alpha_2+\cdots+\alpha_n, \beta). \tag{7.11}$$

The gamma densities with several special values of α and β are shown in Figure 7.1.

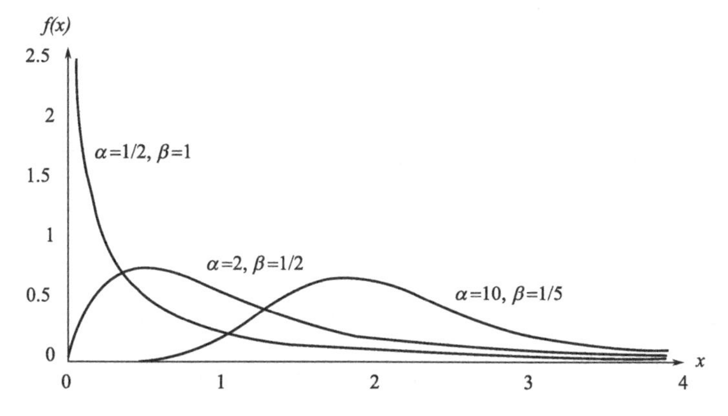

Figure 7.1 Graphs of gamma densities

The exponential distribution that is discussed in Chapter 4 is a special case of gamma distribution when $\alpha=1$ and $\beta=\frac{1}{\lambda}$. We can check that in this case, the probability density of $\Gamma\left(1, \frac{1}{\lambda}\right)$ is

$$f(x)=\begin{cases}\lambda e^{-\lambda x}, & \text{for } x>0, \\ 0, & \text{for } x \leqslant 0.\end{cases}$$

Which is exactly the probability density of an exponential distribution.

Another special case of the gamma distribution, when $\alpha=\frac{v}{2}$ and $\beta=2$, is called Chi-square distribution which also plays an important role in statistics.

Definition 7.7 A random variable X has a Chi-square distribution if its probability density is given by

$$f(x)=\begin{cases}\dfrac{1}{2^{\frac{v}{2}}\Gamma\left(\dfrac{v}{2}\right)}x^{\frac{v-2}{2}}e^{-\frac{x}{2}}, & \text{for } x>0,\\ 0, & \text{for } x\leqslant 0.\end{cases} \tag{7.12}$$

Chi-square distribution can also be written as χ^2 distribution and the parameter v is called the **degree of freedom**[1]. We will write $X\sim\chi^2(v)$ if X is a random variable which follows a Chi-square distribution with the degree of freedom v (see Figure 7.2).

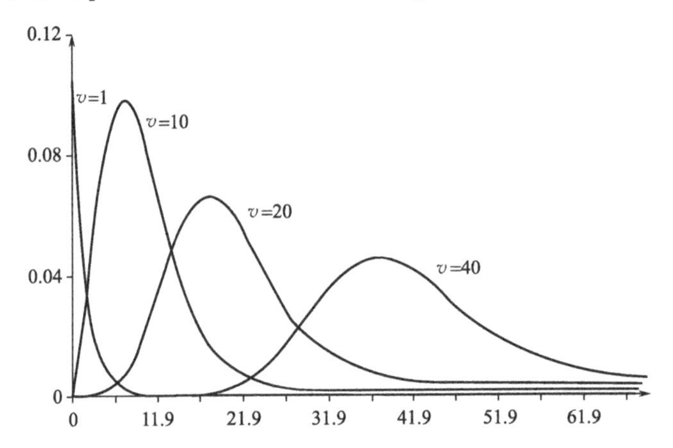

Figure 7.2 Graphs of Chi-square densities

Substituting $\alpha=\dfrac{v}{2}$ and $\beta=2$ into Theorem 7.2, we can get the following corollary.

Corollary 7.2 The mean and the variance of the Chi-square distribution $X\sim\chi^2(v)$ is given by

$$E(X)=v \text{ and } D(X)=2v.$$

The Chi-square distribution is closely related to the normal distribution and has many important applications in statistics. Some of its properties are given below.

Theorem 7.4 Let X has the standard normal distribution $N(0,1)$. Then X^2 follows the Chi-square distribution with degree of freedom $v=1$, in short, $X^2\sim\chi^2(1)$.

With Theorem 7.4, we can prove a more general result.

Theorem 7.5 If $X_1, X_2, \cdots, X_n$ are independent random variables and each of them follows standard normal distribution $N(0,1)$, then

$$\sum_{i=1}^{n} X_i^2 \sim \chi^2(n) \tag{7.13}$$

Proof: Since $X_i\sim N(0,1)$, we obtain, by Theorem 7.4, that

[1] degree of freedom：自由度

$$X_i^2 \sim \chi^2(1) = \Gamma\left(\frac{1}{2}, 2\right), i = 1, 2, \cdots, n.$$

Therefore,

$$\sum_{i=1}^{n} X_i^2 \sim \sum_{i=1}^{n} \Gamma\left(\frac{1}{2}, 2\right) = \Gamma\left(\frac{n}{2}, 2\right) = \chi^2(n)$$

in which we used Theorem 7.3.

Since Chi-square distribution is a special gamma distribution, we are able to obtain further property from Theorem 7.3 that is listed below.

Theorem 7.6 If $X_1, X_2, \cdots, X_n$ are independent random variables and $X_i \sim \chi^2(v_i)$ for $i = 1, 2, \cdots, n$, then

$$X_1 + X_2 + \cdots + X_n \sim \chi^2(v_1 + v_2 + \cdots + v_n). \tag{7.14}$$

Theorem 7.7 Suppose X_1 and X_2 are two independent random variables. If $X_1 \sim \chi^2(v)$ and

$$X_1 + X_2 \sim \chi^2(\mu) \text{ with } \mu > v. \text{ Then } X_2 \sim \chi^2(\mu - v). \tag{7.15}$$

Next, we introduce an important theorem about the distributions of the sample mean and sample variance when the population is a normal distribution.

Theorem 7.8 Suppose $\overline{X}$ and S^2 are the sample mean and the sample variance of a random sample of size n from a population that follows a normal distribution $N(\mu, \sigma^2)$. Then

(1) $\overline{X}$ and S^2 are independent,

(2) $\dfrac{(n-1)S^2}{\sigma^2} \sim \chi^2(n-1)$. (7.16)

Proof: We omit the proof of part (1) since it is beyond the scope of this book. Let us show only part (2).

In order to study the distribution of $\dfrac{(n-1)S^2}{\sigma^2}$, we need the identity

$$\sum_{i=1}^{n} (X_i - \mu)^2 = \sum_{i=1}^{n} (X_i - \overline{X})^2 + n(\overline{X} - \mu)^2. \tag{7.17}$$

In fact, the left of the equation is

$$\begin{aligned}\sum_{i=1}^{n} (X_i - \mu)^2 &= \sum_{i=1}^{n} (X_i^2 - 2X_i\mu + \mu^2) \\ &= \sum_{i=1}^{n} X_i^2 - 2\mu \sum_{i=1}^{n} X_i + n\mu^2 \\ &= \sum_{i=1}^{n} X_i^2 - 2\mu n\overline{X} + n\mu^2\end{aligned}$$

and the right side of the equation is

$$\sum_{i=1}^{n}(X_i-\overline{X})^2+n(\overline{X}-\mu)^2$$
$$=(\sum_{i=1}^{n}X_i^2-2\overline{X}\sum_{i=1}^{n}X_i+n\overline{X}^2)+(n\overline{X}^2-2n\overline{X}\mu+n\mu^2)$$
$$=\sum_{i=1}^{n}X_i^2-2\overline{X}n\overline{X}+2n\overline{X}^2-2n\overline{X}\mu+n\mu^2=\sum_{i=1}^{n}X_i^2-2\mu n\overline{X}+n\mu^2.$$

Which proves (7.17). By the definition of S^2, we have $(n-1)S^2=\sum_{i=1}^{n}(X_i-\overline{X})^2$. Substitute it into (7.17) and then divide both sides by σ^2 to get

$$\sum_{i=1}^{n}\left(\frac{X_i-\mu}{\sigma}\right)^2=\frac{(n-1)S^2}{\sigma^2}+\left(\frac{\overline{X}-\mu}{\sigma/\sqrt{n}}\right)^2.$$

We know that $\frac{X_i-\mu}{\sigma}\sim N(0,1)$ and from Theorem 7.4 and Theorem 7.6, we conclude that $\sum_{i=1}^{n}\left(\frac{X_i-\mu}{\sigma}\right)^2\sim\chi^2(n)$.

At the same time, by Corollary 7.1 and Theorem 7.4 we have $\left(\frac{\overline{X}-\mu}{\sigma/\sqrt{n}}\right)^2\sim\chi^2(1)$. Therefore, it follows by Theorem 7.7 that $\frac{(n-1)S^2}{\sigma^2}\sim\chi^2(n-1)$.

The probability that a random sample produces a $\chi^2(v)$ value greater than some specified value is equal to the area under the curve to the right of this value. It is customary to let $\chi_\alpha^2(v)$ represent the $\chi^2(v)$ above which we find an area of α. That is,

$$P\{\chi^2(v)>\chi_\alpha^2(v)\}=\alpha. \quad (7.18)$$

Figure 7.3 shows the illustration of the $\chi_\alpha^2(v)$.

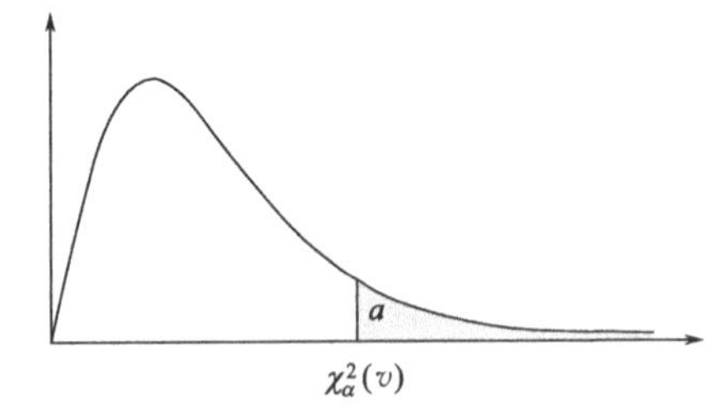

Figure 7.3 Illustration of the $\chi_\alpha^2(v)$

Appendix Table 3 at the end of the book gives the values of $\chi_\alpha^2(v)$ for some selected values of α and v. Hence, the χ^2 value with 7 degrees of freedom, leaving an area of 0.05 to the right, is $\chi_{0.05}^2(7)=14.067$. Owing to lack of symmetry, we must also use the tables to find $\chi_{0.95}^2(7)=2.167$.

7.4.2 Student's Distribution[1] (t-Distribution)

Consider the case that a random sample of size n from a normal population

[1] Student's Distribution：学生氏分布

with mean μ and variance σ^2. In Corollary 7.1, we know that the random variable $\overline{X}$ follows also a normal distribution $N\left(\mu,\frac{\sigma^2}{n}\right)$. Furthermore, $\frac{\overline{X}-\mu}{\sigma/\sqrt{n}} \sim N(0,1)$. In fact, the population standard deviation σ is usually unknown. As a result, a natural statistic to consider is

$$T=\frac{\overline{X}-\mu}{S/\sqrt{n}}.$$

Since S is the sample analog to σ. If the sample size is small, the values of S^2 fluctuate considerably from sample to sample and the distribution of T deviates appreciably from that of a standard normal distribution. In developing the sampling distribution of T, we shall assume that our random sample was selected from a normal population. We can then write

$$T=\frac{(\overline{X}-\mu)/(\sigma/\sqrt{n})}{\sqrt{S^2/\sigma^2}}=\frac{Z}{\sqrt{V/(n-1)}}. \tag{7.19}$$

Where $Z=\frac{\overline{X}-\mu}{\sigma/\sqrt{n}}$ has the standard normal distribution and $V=\frac{(n-1)S^2}{\sigma^2}$ has a chi-squared distribution with $n-1$ degrees of freedom.

In sampling from normal populations, we can show that $\overline{X}$ and S^2 are independent, and consequently so are Z and V. The following theorem gives the definition of a random variable T as a function of Z (standard normal) and χ^2. For completeness, the density function of the t-distribution is given.

Theorem 7.9 Let Z be a standard normal random variable and V a chi-squared random variable with v degrees of freedom. If Z and V are independent, then the distribution of the random variable T, where

$$T=\frac{Z}{\sqrt{V/v}}$$

is given by the density function

$$h(t)=\frac{\Gamma[(v+1)/2]}{\Gamma(v/2)\sqrt{\pi v}}\left(1+\frac{t^2}{v}\right)^{-(v+1)/2}, -\infty<t<+\infty. \tag{7.20}$$

This is known as the **Student t-distribution** with v degrees of freedom.

We omit the detailed proofs which is beyond the requirement of this book. If a random distribution T has the t-distribution with v degrees of freedom, we will write $T\sim t(v)$ for short.

Figure 7.4 depicts the graphs of standard normal distribution, t-distributions with 1 and 5 degrees of freedom. We can see that the curves of t-distributions resemble in general shape the normal distribution. Also, as v increases, the t-distribution will get closer to the normal distribution. In fact, the standard normal distribution can be considered to be the limiting case for the $t(v)$ distributions.

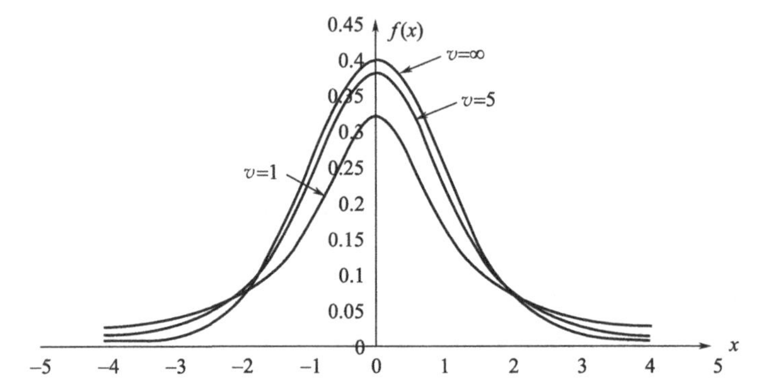

Figure 7.4 The t-distribution curves for $v=1,5$ and $+\infty$

From the foregoing and the theorem above we have the following corollary.

Corollary 7.3 Suppose that $\overline{X}$ and S^2 are the mean and the variance of a random sample of size n from a population which is a normal distribution $N(\mu,\sigma^2)$. Then the random variable $T=\dfrac{\overline{X}-\mu}{S/\sqrt{n}}$ has a t-distribution with $n-1$ degrees of freedom.

In short,

$$T=\frac{\overline{X}-\mu}{S/\sqrt{n}}\sim t(n-1). \tag{7.21}$$

Proof: Let

$$Z=\frac{\overline{X}-\mu}{\sigma/\sqrt{n}} \quad \text{and} \quad Y=\frac{(n-1)S^2}{\sigma^2}$$

By the Corollary 7.1 and the Theorem 7.8, we get that $Z\sim N(0,1)$ and $Y\sim \chi^2(n-1)$, also, Z and Y are independent. Thus,

$$\frac{Z}{\sqrt{V/(n-1)}}=\frac{(\overline{X}-\mu)/(\sigma/\sqrt{n})}{\sqrt{[(n-1)S^2/\sigma_2]/(n-1)}}=\frac{\overline{X}-\mu}{S/\sqrt{n}}.$$

And by Theorem 7.9, we conclude that $T=\dfrac{\overline{X}-\mu}{S/\sqrt{n}}\sim t(n-1)$.

It is customary to let t_α represent the t-value above which we find an area equal to α. Hence, the t-value with 10 degrees of freedom leaving an area of 0.025 to the right is $t_{0.025}(10)=2.228$. Since the t-distribution is symmetric about a mean of zero, we have $t_{1-\alpha}(v)=-t_\alpha(v)$; that is, the t-value leaving an area of $1-\alpha$ to the right and therefore an area of α to the left is equal to the negative t-value that leaves an area of α in the right tail of the distribution (see Figure 7.5). That is, $t_{0.975}(10)=-t_{0.025}(10)=-2.228$.

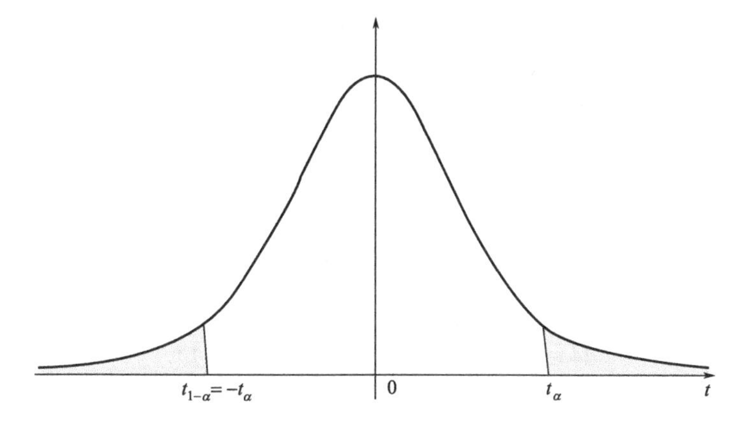

Figure 7.5 Symmetry property of the t-distribution

【Example 7.5】 Find k such that $P(k<T<-1.761)=0.045$ for a random sample of size 15 selected from a normal distribution and $T=\dfrac{\overline{X}-\mu}{S/\sqrt{n}}$.

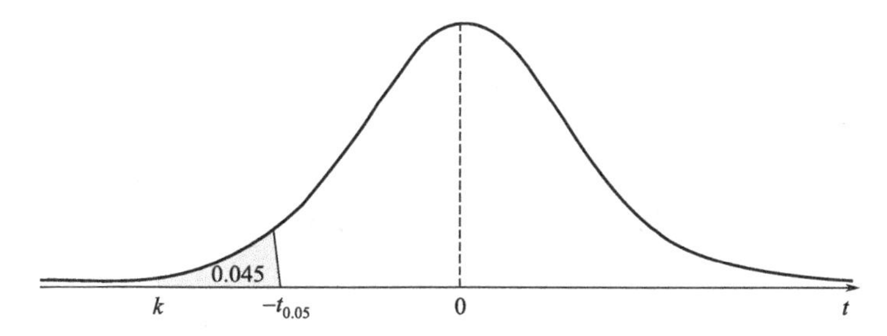

Figure 7.6 The t-values for Example 7.5

Solution: From Corollary 7.3, we know that $T=\dfrac{\overline{X}-\mu}{S/\sqrt{n}}\sim t(14)$. And from appendix Table 4 we note that 1.761 corresponds to $t_{0.05}(14)$. Therefore $-t_{0.05}(14)=-1.761$. Since k in the original probability statement is to the left of $-t_{0.05}(14)=-1.761$, so we let $k=-t_\alpha(14)$. Then, from Figure 7.6, we have $0.045=0.05-\alpha$, or $\alpha=0.005$.

Hence, from appendix Table 4 with $v=14$,

$$k=-t_{0.005}(14)=-2.977$$

and

$$P(-2.977<T<-1.761)=0.045.$$

Corollary 7.4 Let $X_1, X_2, \cdots, X_{n_1}$ and $Y_1, Y_2, \cdots, Y_{n_2}$ are the sample from normal population $N(\mu_1, \sigma_1^2)$ and $N(\mu_2, \sigma_2^2)$ respectively, and they are independent of each other. Suppose $\overline{X}$ and $\overline{Y}$ are the sample mean of the two samples; S_1^2 and S_2^2 are the sample variance of the two samples. Then when $\sigma_1^2=\sigma_2^2=\sigma^2$, we can get

$$\frac{(\overline{X}-\overline{Y})-(\mu_1-\mu_2)}{S_w\sqrt{\frac{1}{n_1}+\frac{1}{n_2}}}\sim t(n_1+n_2-2) \tag{7.22}$$

with
$$S_w^2=\frac{(n_1-1)S_1^2+(n_2-1)S_2^2}{n_1+n_2-2} \tag{7.23}$$

Proof: From Corollary 7.1, we get $\overline{X}-\overline{Y}\sim N\left(\mu_1-\mu_2,\frac{\sigma^2}{n_1}+\frac{\sigma^2}{n_2}\right)$, so

$$U=\frac{(\overline{X}-\overline{Y})-(\mu_1-\mu_2)}{\sigma\sqrt{\frac{1}{n_1}+\frac{1}{n_2}}}\sim N(0,1).$$

And known from Theorem 7.8, we know

$$\frac{(n_1-1)S_1^2}{\sigma^2}\sim\chi^2(n_1-1),\frac{(n_2-1)S_2^2}{\sigma^2}\sim\chi^2(n_2-1)$$

and they are independent. Therefore, it is known by the additivity of the χ^2 distribution that

$$V=\frac{(n_1-1)S_1^2}{\sigma^2}+\frac{(n_2-1)S_2^2}{\sigma^2}\sim\chi^2(n_1+n_2-2).$$

It is also known from Theorem 7.8 that U and V are independent. Then, we can draw the following conclusion from the definition of t-distribution that

$$\frac{U}{\sqrt{V/(n_1+n_2-2)}}=\frac{(\overline{X}-\overline{Y})-(\mu_1-\mu_2)}{S_w\sqrt{\frac{1}{n_1}+\frac{1}{n_2}}}\sim t(n_1+n_2-2).$$

7.4.3 F-distribution[1]

In this section, we will study another important distribution in practical applications of statistics, the F-distribution. While it is of interest to let sampling information shed light on two population means, it is often the case that a comparison of variability is equally important. The F-distribution finds enormous application in comparing sample variances. In order to obtain the F-distribution, we consider the ratio of two independent chi-square random variables, each divided by its own degrees of freedom. The distribution of this ratio is presented in the following theorem.

---

[1] F-distribution: F 分布

Theorem 7.10 Let U and V be two independent random variables having chi-squared distributions with m and n degrees of freedom, respectively. Then the distribution of the random variable $F=\frac{U/m}{V/n}$ has the density function

$$f(x)=\begin{cases}\dfrac{\Gamma\left(\dfrac{m+n}{2}\right)}{\Gamma\left(\dfrac{m}{2}\right)\Gamma\left(\dfrac{n}{2}\right)}m^{\frac{m}{2}}n^{\frac{n}{2}}\dfrac{x^{\frac{m}{2}-1}}{(mx+n)^{\frac{m+n}{2}}}, & \text{if } x>0\\ 0, & \text{if } x\leqslant 0\end{cases} \tag{7.24}$$

This is known as the F-distribution with m and n degrees of freedom (d. f.).

The curve of the F-distribution depends not only on the two parameters m and n but also on the order in which we state them. Once these two values are given, we can identify the curve. Typical F-distributions are shown in Figure 7.7.

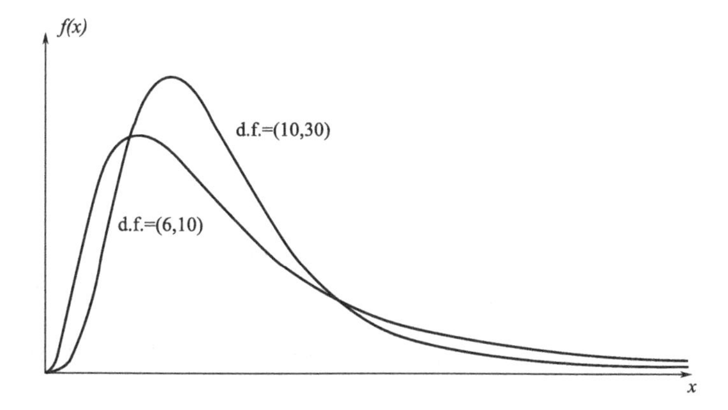

Figure 7.7 Typical F-distributions

Let F_α represent the F-value above which we find an area equal to α. Appendix Table 5 gives values of F_α for $\alpha=0.01$, 0.025, 0.05 and 0.1 for various combinations of the degrees of freedom m and n. Hence, the F-value with 6 and 10 degrees of freedom leaving an area of 0.05 to the right is $F_{0.05}(6,10)=3.22$. By means of the following theorem, Appendix Table 5 can also be used to find values $F_{0.95}$ and $F_{0.99}$.

Theorem 7.11 Writing $F_\alpha(m,n)$ for F_α with m and n degrees of freedom, we obtain

$$F_{1-\alpha}(n,m)=\frac{1}{F_\alpha(m,n)}. \tag{7.25}$$

Proof: Suppose random variable F has F-distribution with n and m degrees of freedom. $F_{1-\alpha}(n,m)$ represent the F-value above which we find an area equal to $1-\alpha$, that is $P\{F>F_{1-\alpha}(n,m)\}=1-\alpha$.

Since $P\{F>F_{1-\alpha}(n,m)\}=P\left\{\frac{1}{F}<\frac{1}{F_{1-\alpha}(n,m)}\right\}$

$$=1-P\left\{\frac{1}{F}\geqslant\frac{1}{F_{1-\alpha}(n,m)}\right\}=1-P\left\{\frac{1}{F}>\frac{1}{F_{1-\alpha}(n,m)}\right\}$$

Then

$$P\left\{\frac{1}{F}>\frac{1}{F_{1-\alpha}(n,m)}\right\}=\alpha. \tag{7.26}$$

By Theorem 7.10, we get that $\frac{1}{F}$ has F-distribution with m and n degrees of freedom, so

$$P\left\{\frac{1}{F}>F_{\alpha}(m,n)\right\}=\alpha. \tag{7.27}$$

From formulas (7.26) and (7.27), we conclude that

$$F_{1-\alpha}(n,m)=\frac{1}{F_{\alpha}(m,n)}.$$

Thus, the F-value with 10 and 6 degrees of freedom, leaving an area of 0.95 to the right, is

$$F_{0.95}(10,6)=\frac{1}{F_{0.05}(6,10)}=\frac{1}{3.22}=0.3106.$$

F-distribution is often applied to the cases when we are interested in comparing the variances of two normal populations or two samples from normal populations. For example, we may need to estimate the ratio $\frac{\sigma_1^2}{\sigma_2^2}$ or to check whether $\sigma_1=\sigma_2$.

Suppose that random samples of size n_1 and n_2 are selected from two normal populations with variances σ_1^2 and σ_2^2. If S_1^2 and S_2^2 are their sample variances, from Theorem 7.8, we know that

$$\frac{(n_1-1)S_1^2}{\sigma_1^2}\sim\chi^2(n_1-1) \quad \text{and} \quad \frac{(n_2-1)S_2^2}{\sigma_2^2}\sim\chi^2(n_2-1).$$

Suppose that these two samples are independent, these two chi-square random variables are also independent. Then, using Theorem 7.10, we obtain the following result.

Theorem 7.12 If S_1^2 and S_2^2 are the variances of independent random samples of size n_1 and n_2 taken from normal populations with variances σ_1^2 and σ_2^2, respectively, then

$$F=\frac{S_1^2/\sigma_1^2}{S_2^2/\sigma_2^2}=\frac{\sigma_2^2S_1^2}{\sigma_1^2S_2^2}\sim F(n_1,n_2). \tag{7.28}$$

Exercises

1. A random sample of employees from a local manufacturing plant pledged the following donations, in dollars, to the United Fund: 100, 40, 75, 15, 20, 100, 75, 50, 30, 10, 55, 75, 25, 50, 90, 80, 15, 25, 45, and 100. Calculate the mean, the mode and the median.
2. The following 10 scores are randomly picked from all the scores in a test of Calculus: 72, 81, 94, 48, 83, 91, 76, 86, 68, 82. Find the sample mean, sample variance and standard deviation.
3. (1) Show that the sample variance is unchanged if a constant c is added to or subtracted from each value in the sample.

   (2) Show that the sample variance becomes c^2 times its original value if each observation in the sample is multiplied by c.
4. A certain type of thread is manufactured with a mean tensile strength of 78.3 kilograms and a standard deviation of 5.6 kilograms. How is the variance of the sample mean changed when the sample size is

   (1) increased from 64 to 196?

   (2) decreased from 784 to 49?
5. If $X_1, X_2, \cdots, X_n$ is a random sample of size n from a Poisson distribution with parameter λ, and $\overline{X}$ is the sample mean. Find $E(\overline{X})$ and $D(\overline{X})$.
6. The heights of 1000 students are normally distributed with a mean of 174.5 centimeters and a standard deviation of 7.9 centimeters. Suppose 200 random samples of size 25 are drawn from this population and the means recorded to the nearest tenth of a centimeter. Find

   (1) the mean and standard deviation of the sampling distribution of $\overline{X}$;

   (2) the number of sample mean that fall between 172.5 and 175.8 centimeters inclusive.
7. Suppose a random sample of size $n=9$ is from a normal distribution with mean $\mu=1$ and standard deviation $\sigma=1.2$.

   (1) Find the mean and standard deviation of the sample mean $\overline{X}$;

   (2) Find the probability that $\overline{X}$ is less than 1.7.
8. If $X_1, X_2, \cdots, X_5$ is a random sample from a normal distribution with mean 12 and variance 4.

   (1) Find $P\{|\overline{X}-\mu|>1\}$;

(2) Find $P\{\max\{X_1, X_2, \cdots, X_5\} > 15\}$ and $P\{\min\{X_1, X_2, \cdots, X_5\} < 10\}$.

9. Given a normal random variable X with mean 20 and variance 9, and a random sample of size n taken from the distribution, what sample size n is necessary in order that $P\{19.9 < \overline{X} < 20.1\} = 0.95$?

10. If independent samples of size 10 and 15 are drawn at random from a normal distribution with mean 20 and variance 3. Find the probability that the absolution of difference of sample means is large than 0.3.

11. If $X_1, X_2, \cdots, X_6$ is a random sample from a standard normal distribution, and random variable $Y = (X_1 + X_2 + X_3)^2 + (X_4 + X_5 + X_6)^2$. Find the value C such that CY has a χ^2 distribution.

12. Let X be a random variable whose distribution is $\chi^2(5)$. Find the value a such that $P\{X \geqslant a\} = 0.05$.

13. Let Y be a t-distribution with degrees of freedom n, find the value a such that
(1) $P\{X > a\} = 0.1$, when $n = 5$;
(2) $P\{X \leqslant a\} = 0.2$, when $n = 8$.

14. The breaking strength X of a certain rivet used in a machine engine has a mean 5000 psi and standard deviation 400 psi. A random sample of 36 rivets is taken. Consider the distribution of $\overline{X}$, the sample mean breaking strength.
(1) What is the probability that the sample mean falls between 4800 psi and 5200 psi?
(2) What sample size n would be necessary in order to have $P\{4900 < \overline{X} < 5100\} = 0.99$?

15. If $X_1, X_2, \cdots, X_6$ is a random sample from a standard normal distribution, and random variable $Y = C(X_1 + X_2)/(X_3^2 + X_4^2 + X_5^2)^{1/2}$. Find the value C such that Y has t-distribution.

16. Show that if T is a random variable which has a t-distribution with n degrees of freedom, then $T^2 \sim F(1, n)$.

17. If S_1^2 and S_2^2 represent the variances of independent random samples of size $n_1 = 8$ and $n_2 = 12$, taken from normal populations with equal variances, find $P\left\{\dfrac{S_1^2}{S_2^2} < 4.89\right\}$.

18. If S_1^2 and S_2^2 represent the variances of independent random samples of size $n_1 = 25$ and $n_2 = 31$, taken from normal populations with variances $\sigma_1^2 = 10$ and $\sigma_2^2 = 15$, find $P\left\{\dfrac{S_1^2}{S_2^2} > 1.26\right\}$.

Chapter 8

Estimation and Uncertainty

The Chapter 7 provided the link between probability and statistical inference. Many statistics are either sums or averages calculated from sample measurements. The Central Limit Theorem states that, even if the sampled populations are not normal, the sampling distributions of these statistics will be approximately normal when the sample size n is large. These statistics are the tools you use for inferential statistics—making inferences about a population using information contained in a sample.

Statistical inference[1] is concerned with making decisions about parameters—the numerical descriptive measures that characterize a population. In statistical inference, a practical problem is restated in the framework of a population with a specific parameter of interest. Methods for making inferences about population parameters fall into one of two areas:

- **Estimation**[2]: Estimating the value of the parameter.
- **Hypothesis testing**[3]: Making a decision about the value of a parameter based on some preconceived idea about what its value might be.

We deal with theory and applications of estimation in this chapter and hypothesis testing in Chapter 9.

8.1 Point Estimation[4]

8.1.1 Some General Concepts of Point Estimation

Suppose that we are interested in learning about the mean height of the

[1] statistical inference：统计推断

[2] estimation：估计

[3] hypothesis testing：假设检验

[4] point estimation：点估计

students of 18 years old in China. One way to obtain the mean would be to measure the height of all students. However, this could be costly, or time-consuming. Instead, we may decide to take a sample of these students, e. g. a sample of 100 students and measure their height. If these 100 students represent well the population, their average value $\bar{x}=175.43$ will be a point estimate of the population mean μ.

In the general, we are interested in estimating some parameter θ of a population X. θ may be a scalar, say the mean μ of the population, a proportion p or a variance σ^2, but θ may be a vector, e. g. both the mean μ and the variance σ^2, written as $\theta=(\mu,\sigma^2)$. In order to carry out this estimation, we usually get a random sample $X_1,X_2,\cdots,X_n$ from population X, which observed values are x_1, $x_2,\cdots,x_n$. The random variable $\overline{X}$ is often used to be an **estimator**[1] of the population mean μ, which observed value $\bar{x}$ is called the **estimate**[2]. More generally, any function of the random sample $X_1,X_2,\cdots,X_n$ used to estimate θ is called an estimator of θ and is denoted by $\hat{\theta}$. In the case of the above example $\hat{\theta}=\hat{\mu}=\overline{X}$. Summarizing the above we have the definition.

Definition 8.1 Suppose θ is a parameter of a population, $X_1,X_2,\cdots,X_n$ is a random sample from this population, and $T(X_1,X_2,\cdots,X_n)$ is a statistic that is a function of $X_1,X_2,\cdots,X_n$. If we use the observed value $T(x_1,x_2,\cdots,x_n)$ to estimate parameter θ, $T(X_1,X_2,\cdots,X_n)$ is called **a point estimator** of θ and $T(x_1,x_2,\cdots,x_n)$ is called as **a point estimate** of θ.

In the height of the students example, the estimator used to obtain the point estimate of μ is $\overline{X}$, and the point estimate of μ is 175.43. The symbol $\hat{\theta}$ is customarily used to denote both the estimator of θ and the point estimate resulting from a given sample.

【Example 8.1】 The accompanying sample consisting of n=20 observations on dielectric breakdown voltage of a piece of epoxy resin.

Observation:

24.46, 25.61, 26.25, 26.42, 26.66, 27.15, 27.31, 27.54, 27.74, 27.94
27.98, 28.04, 28.28, 28.49, 28.50, 28.87, 29.11, 29.13, 29.50, 30.88

We assume that the distribution of breakdown voltage is normal with mean value μ. Because normal distributions are symmetric, μ is also the median lifetime of the distribution. The given observations are then assumed to be the result of a random

[1] estimator：估计量
[2] estimate：估计值

sample $X_1, X_2, \cdots, X_{20}$ from this normal distribution. Consider the following estimators and resulting estimates for μ.

(1) Estimator $\hat{\mu}=\overline{X}$, estimate $\hat{\mu}=\overline{x}=(\sum_{i=1}^{20} x_i)/20=555.86/20=27.793$.

(2) Estimator $\hat{\mu}=\widetilde{X}$, estimate $\hat{\mu}=\widetilde{x}=(27.94+27.98)/2=27.960$.

(3) Estimator $\hat{\mu}=[\min(X_i)+\max(X_i)]/2$, estimate

$$\hat{\mu}=[\min(x_i)+\max(x_i)]/2=(24.46+30.88)/2=27.670.$$

【Example 8.2】 Consider the following sample of observations on elastic modulus (GPa) of AZ91D alloy specimens from a die-casting process: 44.2 43.9 44.7 44.2 44.0 43.8 44.6 43.1.

We want to estimate the population variance σ^2. A natural estimator is the sample variance:

$$\hat{\sigma}^2=S^2=\frac{\sum_{i=1}^{n}(X_i-\overline{X})^2}{n-1}=\frac{\sum_{i=1}^{n}X_i^2-(\sum_{i=1}^{n}X_i)^2/n}{n-1}.$$

Thus, the point estimate of the population variance σ^2 is

$$\hat{\sigma}^2=s^2=\frac{\sum_{i=1}^{8}x_i^2-(\sum_{i=1}^{8}x_i)^2/8}{8-1}=\frac{15533.79-(352.5)^2/8}{7}=0.25125.$$

The point estimate of σ would then be $\hat{\sigma}=s=\sqrt{0.25125}=0.501$.

An alternative estimator results from using the divisor n rather than $n-1$:

$$\hat{\sigma}^2=\frac{\sum_{i=1}^{n}(X_i-\overline{X})^2}{n}.$$

Estimate $\hat{\sigma}^2=\frac{1.75875}{8}=0.220$.

We will shortly indicate why many statisticians prefer S^2 to this latter estimator.

8.1.2 Selection Criteria of Point Estimators

1. Unbiased Estimator[1]

In a practical situation, there may be several statistics that could be used as point estimators for a population parameter. To decide which of several choices is best, you need to know how the estimator behaves in repeated sampling, de-

[1] unbiased estimator：无偏估计

scribed by its sampling distribution.

Sampling distributions provide information that can be used to select the best estimator. What characteristics would be valuable? First, the sampling distribution of the point estimator should be centered over the true value of the parameter to be estimated. That is, the estimator should not constantly underestimate or overestimate the parameter of interest. Such an estimator is said to be unbiased.

Definition 8.2 A point estimator $\hat{\theta}$ is said to be an **unbiased estimator** of θ if

$$E(\hat{\theta})=\theta. \tag{8.1}$$

Otherwise, the estimator is said to be biased, the difference $E(\hat{\theta})-\theta$ is called the bias of $\hat{\theta}$.

The distributions for biased and unbiased estimators are shown in Figure 8.1.

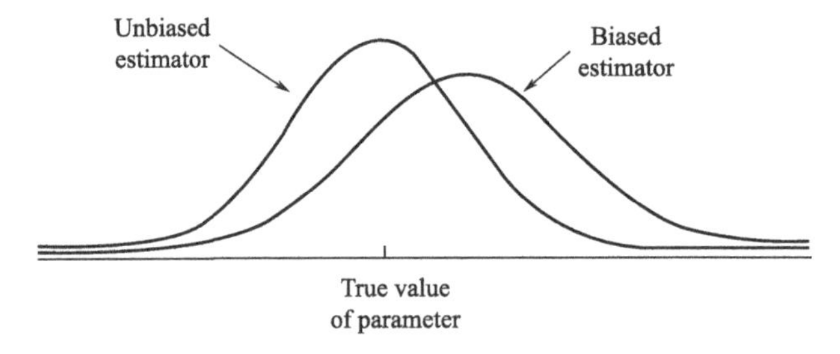

Figure 8.1 Distributions for biased and unbiased estimators

【Example 8.3】 If $\overline{X}$ is the sample mean of a random sample of size n from a Poisson population X with parameter λ, show that $\overline{X}$ is an unbiased estimator of λ.

Proof: Since the population X has Poisson distribution, then $E(X)=\lambda$.

Therefore $E(\overline{X})=E(X)=\lambda$.

So $\overline{X}$ is an unbiased estimator of λ.

The estimator $\overline{X}$ is unbiased for estimating population mean μ.

【Example 8.4】 Let $X_1, X_2, \cdots, X_n$ be a random sample from a distribution with mean μ and variance σ^2. Then the estimator $\hat{\sigma}^2=S^2=\frac{1}{n-1}\sum_{i=1}^{n}(X_i-\overline{X})^2$ is unbiased for estimating σ^2.

Proof: For any random variable X, $Var(X)=E(X^2)-[E(X)]^2$, then

$$E(X^2)=Var(X)+[E(X)]^2.$$

Therefore $$E(X_i^2)=\sigma^2+\mu^2 \text{ for } i=1,2,\cdots,n.$$

Applying this to $S^2=\frac{1}{n-1}\sum_{i=1}^{n}(X_i-\overline{X})^2=\frac{1}{n-1}\left[\sum_{i=1}^{n}X_i^2-\frac{1}{n}\left(\sum_{i=1}^{n}X_i\right)^2\right]$, gives

$$E(S^2)=\frac{1}{n-1}\left[\sum_{i=1}^{n}E(X_i^2)-\frac{1}{n}E\left(\sum_{i=1}^{n}X_i\right)^2\right]$$

$$=\frac{1}{n-1}\left\{\sum_{i=1}^{n}(\sigma^2+\mu^2)-\frac{1}{n}\left\{Var\left(\sum_{i=1}^{n}X_i\right)+\left[E\left(\sum_{i=1}^{n}X_i\right)\right]^2\right\}\right\}$$

$$=\frac{1}{n-1}\left[n\sigma^2+n\mu^2-\frac{1}{n}n\sigma^2-\frac{1}{n}(n\mu)^2\right]$$

$$=\frac{1}{n-1}(n\sigma^2-\sigma^2)=\sigma^2.$$

Therefore, the estimator $S^2=\frac{1}{n-1}\sum_{i=1}^{n}(X_i-\overline{X})^2$ is an unbiased estimator for σ^2.

Sometimes, the estimator is not unbiased, but the expected value of the estimator will approach to the parameter when the sample size n is large. Such estimators are called **asymptotically unbiased estimators**[1].

Definition 8.3 A point estimator $\hat{\theta}_n$ is said to be an **asymptotically unbiased estimator** of θ, if

$$\lim_{n\to\infty}E(\hat{\theta}_n)=\theta. \tag{8.2}$$

For example, the estimator $\hat{\sigma}^2=\frac{1}{n}\sum_{i=1}^{n}(X_i-\overline{X})^2$ for parameter variance σ^2 is not unbiased because of $E(\hat{\sigma}^2)=E\left(\frac{n-1}{n}S^2\right)=\frac{n-1}{n}\sigma^2$, but $\lim_{n\to\infty}E(\hat{\sigma}^2)=\sigma^2$. Therefore, $\hat{\sigma}^2=\frac{1}{n}\sum_{i=1}^{n}(X_i-\overline{X})^2$ is an asymptotically unbiased estimator for σ^2.

When choosing among several different estimators of parameter θ, select one that is unbiased is our principle. According to this principle, the unbiased estimator $S^2=\frac{1}{n-1}\sum_{i=1}^{n}(X_i-\overline{X})^2$ in Example 8.4 should be preferred to the biased estimator $\frac{1}{n}\sum_{i=1}^{n}(X_i-\overline{X})^2$.

In Example 8.1, we proposed several different estimators for the mean μ of a normal distribution. If there were a unique unbiased estimator for μ. the estimation problem would be resolved by using that estimator. Unfortunately, this is not the case. The estimator $\overline{X}$ is an unbiased estimator of μ. If in addition the distri-

---

[1] asymptotically unbiased estimators：渐近无偏估计量

bution is continuous and symmetric, then $\widetilde{X}$ and any trimmed mean are also unbiased estimators of μ. So, the principle of unbiasedness by itself does not always allow us to select a single estimator. What we now need is a way of selecting among unbiased estimators.

2. Estimators with Minimum Variance

It is very possible that we have more than one unbiased estimator for a given parameter θ. Although the distribution of each estimator is centered at the true value of θ, the spreads of the distributions about the true value may be different. Usually we prefer the one with smallest variance. This ensures that, with a high probability, an individual estimate will fall close to the true value of θ. The sampling distributions for two unbiased estimators, one with a small variance and the other with a larger variance, are shown in Figure 8.2. Naturally, you would prefer the estimator with the smaller variance because the estimates tend to lie closer to the true value of the parameter than in the distribution with the larger variance. The unbiased estimator for a parameter with the smallest variance among all unbiased estimators for this parameter is called **minimum variance unbiased estimator**[1].

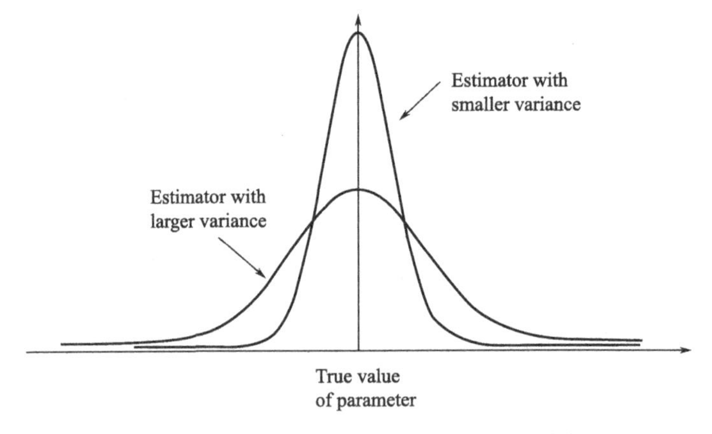

Figure 8.2 Comparison of estimator variability

Definition 8.4 Let $\hat{\theta},\hat{\theta}^*$ be unbiased estimators of θ. If $D(\hat{\theta}^*)\leqslant D(\hat{\theta})$, we call $\hat{\theta}^*$ is more efficiency than $\hat{\theta}$. If for any $\hat{\theta}'$ which is also an unbiased estimator of θ, we have $D(\hat{\theta})\leqslant D(\hat{\theta}')$, Then $\hat{\theta}$ is called the **minimum variance unbiased estimator (MVUE)** of θ.

【Example 8.5】 Let $X_1, X_2, \cdots, X_n$ be a random sample from population X with

[1] minimum variance unbiased estimator：最小方差无偏估计量

mean μ. $\overline{X}=\frac{1}{n}\sum_{i=1}^{n}X_i$ and $\overline{W}=\sum_{i=1}^{n}a_iX_i$ ($a_i\geqslant 0$, and $\sum_{i=1}^{n}a_i=1$) are point estimator for μ. Prove that $\overline{X}$ and $\overline{W}$ are all unbiased estimators of μ, and $\overline{X}$ is more efficiency than $\overline{W}$ when $a_1, a_2, \cdots a_n$ are unequal.

Proof: Since $E(X_i)=\mu$ for any i, then

$$E(\overline{X})=\frac{1}{n}\sum_{i=1}^{n}E(X_i)=\mu,$$

and

$$E(\overline{W})=\sum_{i=1}^{n}a_iE(X_i)=\mu\sum_{i=1}^{n}a_i=\mu.$$

Therefore, $\overline{X}$ and $\overline{W}$ are all unbiased estimators of μ.

$$D(\overline{X})=\frac{1}{n^2}\sum_{i=1}^{n}D(X_i)=\frac{1}{n}D(X),$$

$$D(\overline{W})=\sum_{i=1}^{n}a_i^2D(X_i)=D(X)\sum_{i=1}^{n}a_i^2.$$

Since $\sum_{i=1}^{n}a_i^2>\frac{1}{n}(\sum_{i=1}^{n}a_i)^2=\frac{1}{n}$ when $a_1, a_2, \cdots a_n$ are unequal, then $D(\overline{W})>D(\overline{X})$.

So, $\overline{X}$ is more efficiency than $\overline{W}$.

To judge the efficiency of an unbiased estimator, the Cramer-Rao Inequality, given in formula (8.3) below, is often used.

$$D(\hat{\theta})\geqslant\frac{1}{nE\left[\left(\frac{\partial \ln f(X)}{\partial\theta}\right)^2\right]} \tag{8.3}$$

Where $f(x)$ is the population density function and n is the sample size. The proof of this inequality is omitted.

The Cramer-Rao inequality leads to the following theorem.

Theorem 8.1 Let $\hat{\theta}$ be an unbiased estimator of θ. If

$$D(\hat{\theta})=\frac{1}{nE\left[\left(\frac{\partial \ln f(X)}{\partial\theta}\right)^2\right]} \tag{8.4}$$

Then $\hat{\theta}$ is a minimum variance unbiased estimator of θ.

【Example 8.6】 Suppose that $\overline{X}$ is the sample mean of a random sample of size n from a normal population $N(\mu,\sigma^2)$. Show that $\overline{X}$ is a minimum variance unbiased estimator of μ.

Proof: The population density function is

$$f(x)=\frac{1}{\sqrt{2\pi}\sigma}e^{-\frac{(x-\mu)^2}{2\sigma^2}}.$$

Which leads to

$$\frac{\partial \ln f(x)}{\partial \mu}=\frac{\partial}{\partial \mu}\left[-\ln(\sigma\sqrt{2\pi})-\frac{1}{2}\left(\frac{x-\mu}{\sigma}\right)^2\right]=\frac{1}{\sigma}\left(\frac{x-\mu}{\sigma}\right).$$

Therefore,

$$E\left[\left(\frac{\partial \ln f(X)}{\partial \mu}\right)^2\right]=E\left[\frac{1}{\sigma^2}\left(\frac{X-\mu}{\sigma}\right)^2\right]=\frac{1}{\sigma^2}E\left[\left(\frac{X-\mu}{\sigma}\right)^2\right]=\frac{1}{\sigma^2}.$$

Thus,

$$\frac{1}{nE\left[\left(\frac{\partial \ln f(X)}{\partial \mu}\right)^2\right]}=\frac{1}{n\frac{1}{\sigma^2}}=\frac{\sigma^2}{n}=D(\overline{X}).$$

As we know, $E(\overline{X})=\mu$. By Theorem 8.1, we conclude that $\overline{X}$ is a minimum variance unbiased estimator of μ.

From the above discussion we have **the basic rule of a good estimator that it must be unbiased with small variance.** For example, if two estimators $\hat{\theta}_1,\hat{\theta}_2$ are both unbiased, then we would choose the one with the smaller variance. If $\hat{\theta}_1,\hat{\theta}_2$ are not unbiased, the two estimators are not comparable, i. e. there is no point in looking at smaller variance.

3. Consistent estimator[1]

Another property of a good estimator is that of consistency. Suppose we have an estimator $\hat{\theta}_n$ of θ, $\hat{\theta}_n$ is a consistent estimator if it converges in probability to θ, as n goes to infinity. In other words, increasing n increases the probability of $\hat{\theta}_n$ being close to θ.

Definition 8.5 The estimator $\hat{\theta}_n$ of θ is called **consistent estimator** if

$$\lim_{n\to\infty}P\{|\hat{\theta}_n-\theta|<\varepsilon\}=1 \tag{8.5}$$

or equivalently, if

$$\lim_{n\to\infty}P\{|\hat{\theta}_n-\theta|\geqslant\varepsilon\}=0 \tag{8.6}$$

for any small $\varepsilon>0$.

For example, for the estimator $\hat{\mu}=\overline{X}$ of μ in Example 8.5, we can prove that $\lim\limits_{n\to\infty}P\{|\overline{X}-\mu|\geqslant\varepsilon\}=0$, so $\overline{X}$ is unbiased and consistent estimator for μ.

[1] consistent estimator：一致估计量

We note that the above example indicate that an unbiased estimator is consistent too. There is an easier way to check whether unbiased estimators are consistent and this is given in the following theorem.

Theorem 8.2 Suppose that $\hat{\theta}_n$ is an unbiased estimator of θ. If

$$\lim_{n\to\infty} D(\hat{\theta}_n)=0, \tag{8.7}$$

then $\hat{\theta}_n$ is consistent.

Proof: From Chebychev's inequality, we know $P\{|\hat{\theta}_n-E(\hat{\theta}_n)|\geqslant\varepsilon\}\leqslant\frac{D(\hat{\theta}_n)}{\varepsilon^2}$.

Since $\lim\limits_{n\to\infty} D(\hat{\theta}_n)=0$, then $\lim\limits_{n\to\infty} P\{|\hat{\theta}_n-\theta|\geqslant\varepsilon\}=\lim\limits_{n\to\infty} P\{|\hat{\theta}_n-E(\hat{\theta}_n)|\geqslant\varepsilon\}=0$.

Thus, $\hat{\theta}_n$ is consistent estimator of θ.

This theorem enables us to check quickly whether an unbiased estimator is consistent, without needing to evaluate the limit of $P(|\hat{\theta}_n-\theta|\geqslant\varepsilon)$. We note that in the case of the $\overline{X}$ estimator, applying Theorem 8.2, would involve us noting that $\overline{X}$ is unbiased for the mean μ and that its variance σ^2/n converges to 0 as $n\to\infty$.

But not all estimators are unbiased and not all estimators are consistent. In fact, a biased estimator may be consistent and an unbiased estimator may be inconsistent. Below we give an example of each case.

【Example 8.7】 Suppose $X_1, X_2, \cdots, X_n$. be a random sample from normal population with mean μ and variance σ^2.

(1) Prove estimator $\hat{\mu}_n=\overline{X}+\frac{1}{n}$ is biased and consistent.

(2) Prove estimator $\hat{\mu}_n=X_1$ is unbiased, but not consistent.

Proof: (1) Since $E(\hat{\mu}_n)=E\left(\overline{X}+\frac{1}{n}\right)=\mu+\frac{1}{n}\neq\mu$, then $\hat{\mu}_n=\overline{X}+\frac{1}{n}$ is biased estimator for μ.

By Corollary 7.1 we can get $\overline{X}\sim N\left(\mu,\frac{\sigma^2}{n}\right)$, so $\hat{\mu}_n\sim N\left(\mu+\frac{1}{n},\frac{\sigma^2}{n}\right)$.

Therefore $P\{|\hat{\mu}_n-\mu|\geqslant\varepsilon\}=P\{\hat{\mu}_n-\mu\geqslant\varepsilon\}+P\{\hat{\mu}_n-\mu\leqslant-\varepsilon\}$

$$=1-P\left\{\frac{\hat{\mu}_n-\mu-\frac{1}{n}}{\sigma/\sqrt{n}}<\frac{\varepsilon-\frac{1}{n}}{\sigma/\sqrt{n}}\right\}+P\left\{\frac{\hat{\mu}_n-\mu-\frac{1}{n}}{\sigma/\sqrt{n}}\leqslant\frac{-\varepsilon-\frac{1}{n}}{\sigma/\sqrt{n}}\right\}$$

$$=1-\Phi\left(\frac{\varepsilon-\frac{1}{n}}{\sigma/\sqrt{n}}\right)+\Phi\left(\frac{-\varepsilon-\frac{1}{n}}{\sigma/\sqrt{n}}\right).$$

Since $\frac{\varepsilon-\frac{1}{n}}{\sigma/\sqrt{n}}\to+\infty$ and $\frac{-\varepsilon-\frac{1}{n}}{\sigma/\sqrt{n}}\to-\infty$ when $n\to\infty$, then

$$\lim_{n\to\infty}\Phi\left(\frac{\varepsilon-\frac{1}{n}}{\sigma/\sqrt{n}}\right)=1,\ \lim_{n\to\infty}\Phi\left(\frac{-\varepsilon-\frac{1}{n}}{\sigma/\sqrt{n}}\right)=0.$$

Thus
$$\lim_{n\to\infty}P\{|\hat{\mu}_n-\mu|\geqslant\varepsilon\}=0.$$

Therefore, $\hat{\mu}_n=\overline{X}+\frac{1}{n}$ is a consistent estimator for μ.

(2) Since $X_1\sim N(\mu,\sigma^2)$, then $E(\hat{\mu}_n)=E(X_1)=\mu$, so $\hat{\mu}_n=X_1$ is unbiased estimator for μ.

$$P\{|X_1-\mu|\geqslant\varepsilon\}=P\{X_1-\mu\geqslant\varepsilon\}+P\{X_1-\mu\leqslant-\varepsilon\}$$
$$=1-P\{X_1-\mu\leqslant\varepsilon\}+P\{X_1-\mu\leqslant-\varepsilon\}=1-P\left\{\frac{X_1-\mu}{\sigma}\leqslant\frac{\varepsilon}{\sigma}\right\}+P\left\{\frac{X_1-\mu}{\sigma}\leqslant-\frac{\varepsilon}{\sigma}\right\}$$
$$=1-\Phi\left(\frac{\varepsilon}{\sigma}\right)+\Phi\left(-\frac{\varepsilon}{\sigma}\right)=2-2\Phi\left(\frac{\varepsilon}{\sigma}\right).$$
$$\lim_{n\to\infty}P\{|X_1-\mu|\geqslant\varepsilon\}\neq0.$$

Thus, $\hat{\mu}_n=X_1$ is unbiased, but not consistent estimator for μ.

8.2 Method of Point Estimation

We now introduce two "constructive" methods for obtaining point estimators: the method of moments and the method of maximum likelihood. By constructive we mean that the general definition of each type of estimator suggests explicitly how to obtain the estimator in any specific problem. Although maximum likelihood estimators are generally preferable to moment estimators because of certain efficiency properties, they often require significantly more computation than do moment estimators. It is sometimes the case that these methods yield unbiased estimators.

8.2.1 Method of Moments[1]

Method of moment, which was first proposed by Karl Pearson, is one of the oldest methods. It is simple and easy to use in many problems. The brief idea of the method of moments is to equate the moments of a population to the correspond-

[1] method of moments：矩估计法

ing moments of a random sample. Then solving these equations for unknown parameters yields the estimators.

Definition 8.6 Let $X_1, X_2, \cdots, X_n$ be a random sample from a population with k unknown parameters $\theta_1, \theta_2, \cdots, \theta_k$. Then the moment estimators $\hat{\theta}_1, \hat{\theta}_2, \cdots, \hat{\theta}_k$ are obtained by equating the first k sample moments to the corresponding first k population moments and solving for $\theta_1, \theta_2, \cdots, \theta_k$.

For example, if $k=2$, $E(X)$ and $E(X^2)$ will be functions of θ_1 and θ_2. Setting $E(X)=\overline{X}$ and $E(X^2)=\frac{1}{n}\sum_{i=1}^{n}X_i^2$ gives two equations in θ_1 and θ_2. The solution then defines the estimators $\hat{\theta}_1$ and $\hat{\theta}_2$.

【Example 8.8】 Let $X_1, X_2, \cdots, X_n$ be a random sample from an exponential population X with parameter λ. Find an estimator of the parameter λ by method of moments.

Solution: This problem has only one unknown parameter to be estimated. So, we only need to set up the first equation $EX=\overline{X}$. Since $EX=\frac{1}{\lambda}$ for an exponential distribution, then the equation becomes

$$\frac{1}{\lambda}=\overline{X}.$$

The moment estimator of λ is then $\hat{\lambda}=\frac{1}{\overline{X}}$.

【Example 8.9】 Suppose X is a population with $E(X)=\mu$ and $D(X)=\sigma^2$. Find the estimators for μ and σ^2 by method of moments.

Solution: Let $X_1, X_2, \cdots, X_n$ be a random sample from population X.

We set up two equations

$$\begin{cases} E(X)=\overline{X}, \\ E(X^2)=\frac{1}{n}\sum_{i=1}^{n}X_i^2. \end{cases}$$

Since $E(X^2)=D(X)+[E(X)]^2$ and $B_2=\frac{1}{n}\sum_{i=1}^{n}(X_i-\overline{X})^2$, the above equations are equivalent to

$$\begin{cases} \mu=\overline{X}, \\ \sigma^2+\mu^2=\frac{1}{n}\sum_{i=1}^{n}X_i^2. \end{cases}$$

We conclude that

$$\begin{cases}\hat{\mu}=\overline{X},\\ \hat{\sigma}^2=\frac{1}{n}\sum_{i=1}^{n}X_i^2-(\overline{X})^2=B_2,\end{cases}$$

are the estimators for μ and σ^2 by method of moments.

From the above example we know that when we have more than one unknown parameters, the second moment equation $E(X^2)=\frac{1}{n}\sum_{i=1}^{n}X_i^2$ is equivalent to $D(X)=B_2$.

【Example 8.10】 Let $X_1, X_2, \cdots, X_n$ be a random sample from a uniform population X on $[\alpha, \beta]$. Find estimators of the parameter α and β by method of moments.

Solution: We know from Chapter 5 that $E(X)=\frac{\alpha+\beta}{2}$ and $D(X)=\frac{(\beta-\alpha)^2}{12}$.

We set up two equations

$$\begin{cases}E(X)=\overline{X},\\ D(X)=B_2.\end{cases}$$

That is

$$\begin{cases}\frac{\alpha+\beta}{2}=\overline{X},\\ \frac{(\beta-\alpha)^2}{12}=B_2.\end{cases}$$

Solve the equations and we obtain

$$\begin{cases}\hat{\alpha}=\overline{X}-\sqrt{3B_2},\\ \hat{\beta}=\overline{X}+\sqrt{3B_2}.\end{cases}$$

Which are two estimators of α and β by method of moments.

8.2.2 Method of Maximum Likelihood[1]

The method of maximum likelihood which is the most important methods of point estimation was first introduced by R. A. Fisher, a geneticist and statistician, in the 1920s. Most statisticians recommend this method, at least when the sample size is large, since the resulting estimators have certain desirable efficiency properties.

This method is based on the idea that the best choice among all the possible estimations for a parameter is the one which an observed result has the greatest probability.

For example, a box has 5 balls inside it. The color of the balls is red and yellow, but we don't know any information about the number of red balls. If we

---

[1] method of maximum likelihood：最大似然估计

pick up two balls randomly and find that they are both red ball, what is the most reasonable number of the red balls?

We suppose there are $k(k=0,1,2,3,4,5)$ red balls in the box. Then, for any given k, the probability of picking up 2 red balls is

$$P_k=\frac{C_k^2C_{5-k}^0}{C_5^2},k=0,1,2,3,4,5.$$

All the results of P_k are listed in the following table.

k	0	1	2	3	4	5
P_k	0	0	0.1	0.3	0.6	1

From the above table we can see that $k=5$ makes the largest probability of picking up 2 red balls, so that 5 is the estimate of the red balls. $k=5$ is called the maximum likelihood estimate of the number of red balls and this method is called the method of maximum likelihood.

In discrete case, suppose $X_1,X_2,\cdots,X_n$ be a random sample from a discrete population X which probability mass function is $P\{X=x_j\}=p(x_j;\theta)(j=1,2,\cdots)$ with an unknown parameter θ, and the sample values are $x_1,x_2,\cdots,x_n$. Then the probability of getting these values is

$$P\{X_1=x_1,X_2=x_2,\cdots,X_n=x_n\}=\prod_{i=1}^{n}p(x_i;\theta) \qquad (8.8)$$

Which is the joint probability distribution of the random variables $X_1,X_2,\cdots,X_n$ at the values $x_1,x_2,\cdots,x_n$. Since the sample values are known, $\prod_{i=1}^{n}p(x_i;\theta)$ is a function of θ.

Definition 8.7 A random sample has the observed values $x_1,x_2,\cdots,x_n$ from a discrete population with an unknown parameter θ. The function defined in (8.8) is called **the likelihood function**[1]. Usually we denote it as $L(\theta)$, that is

$$L(\theta)=\prod_{i=1}^{n}p(x_i;\theta). \qquad (8.9)$$

In the continuous case, suppose $X_1,X_2,\cdots,X_n$ be a random sample from a population X which probability density function is $f(x,\theta)$. Then the probability of $(X_1,X_2,\cdots,X_n)$ falls into the neighborhood of the observation value $(x_1,x_2,\cdots,$

[1] likelihood function：似然函数

x_n) will be determined by $\prod_{i=1}^{n} f(x_i;\theta)$. Therefore, we define the likelihood function

$$L(\theta)=\prod_{i=1}^{n} f(x_i;\theta). \qquad (8.10)$$

The task of the maximum likelihood method is to find a value θ by maximize the likelihood function. Such value is called a maximum likelihood estimate of θ.

Definition 8.8 Let $X_1, X_2, \cdots, X_n$ be a random sample from a population X, and the observed values are $x_1, x_2, \cdots, x_n$. If there is a value $\hat{\theta}=\hat{\theta}(x_1, x_2, \cdots, x_n)$ such that $L(\hat{\theta}) \geqslant L(\theta)$ for all θ, then $\hat{\theta}(x_1, x_2, \cdots, x_n)$ is called a **maximum likelihood estimate**[1] of θ, the corresponding $\hat{\theta}(X_1, X_2, \cdots, X_n)$ is called a **maximum likelihood estimator**[2] of θ.

We know that a necessary condition of $L(\hat{\theta})$ being a maximum value is that $\left.\frac{\mathrm{d}L(\theta)}{\mathrm{d}\theta}\right|_{\theta=\hat{\theta}}=0$. The equation $\frac{\mathrm{d}\ln L(\theta)}{\mathrm{d}\theta}=0$ is called **likelihood equation**[3]. Since $\ln x$ is a strictly increasing function in its domain, then both $L(\theta)$ and $\ln L(\theta)$ reach the maximum value simultaneously. In many cases, it is easier to find the maximum value of $\ln L(\theta)$ than $L(\theta)$. So, we often solve the equation $\frac{\mathrm{d}\ln L(\theta)}{\mathrm{d}\theta}=0$ to find the maximum likelihood estimate of θ. The equation $\frac{\mathrm{d}\ln L(\theta)}{\mathrm{d}\theta}=0$ is called **Log-likelihood equation**[4].

【Example 8.11】 Let $x_1, x_2, \cdots, x_n$ be the values of a random sample from a Poisson distribution $p(x;\lambda)=\frac{\lambda^x}{x!}\mathrm{e}^{-\lambda}$ with $\lambda>0$ and $x=0,1,2,\cdots$. Find the maximum likelihood estimator of parameter λ.

Solution: The likelihood function is

$$L(\lambda)=\prod_{i=1}^{n} p(x_i;\lambda)=\prod_{i=1}^{n} \frac{\lambda^{x_i}}{x_i!}\mathrm{e}^{-\lambda}=\frac{\lambda^{\sum_{i=1}^{n} x_i}}{\prod_{i=1}^{n}(x_i!)}\mathrm{e}^{-n\lambda}.$$

---

[1] maximum likelihood estimate：最大似然估计值
[2] maximum likelihood estimator：最大似然估计量
[3] likelihood equation：似然方程
[4] Log-likelihood equation：对数似然方程

and so

$$\ln L(\lambda) = (\sum_{i=1}^{n} x_i) \cdot \ln \lambda - \sum_{i=1}^{n} \ln(x_i !) - n\lambda.$$

Set the differentiation of $\ln L(\lambda)$ with respect to λ equal to 0 to get

$$\frac{d}{d\lambda} \ln L(\lambda) = \frac{1}{\lambda}(\sum_{i=1}^{n} x_i) - n = 0.$$

The solution of the above equation is

$$\hat{\lambda} = \frac{1}{n}(\sum_{i=1}^{n} x_i) = \bar{x}.$$

Therefore, the maximum likelihood estimator for parameter λ is $\hat{\lambda} = \bar{X}$.

【Example 8. 12】 Let $X_1, X_2, \cdots, X_n$ be a random sample from a population with probability density function

$$f(x;\theta) = \frac{1}{2\theta} e^{-\frac{|x|}{\theta}}, -\infty < x < +\infty.$$

Find the estimator of θ by method of maximum likelihood.

Solution: The likelihood function is

$$L(\theta) = \prod_{i=1}^{n} f(x_i;\theta) = \frac{1}{(2\theta)^n} e^{-\sum_{i=1}^{n} \frac{|x_i|}{\theta}},$$

$$\ln L(\theta) = -\frac{1}{\theta} \sum_{i=1}^{n} |x_i| - n\ln(2\theta).$$

Set

$$\frac{d \ln L(\theta)}{d\theta} = \frac{1}{\theta^2} \sum_{i=1}^{n} |x_i| - \frac{n}{\theta} = 0.$$

The solution of the above equation is

$$\hat{\theta} = \frac{1}{n} \sum_{i=1}^{n} |x_i|.$$

Therefore, the maximum likelihood estimator for parameter θ is

$$\hat{\theta} = \frac{1}{n} \sum_{i=1}^{n} |X_i|.$$

The method of maximum likelihood can also be applied to the problems in which there are more than one unknown parameters. In those cases, the likelihood function will be $L(\theta_1, \theta_2, \cdots, \theta_m)$, where $\theta_1, \theta_2, \cdots, \theta_m$ are m unknown parameters. In order to find the maximum value of the likelihood function, a common method is to solve the system of equations

$$\begin{cases}\frac{\partial}{\partial\theta_1}\ln L(\theta_1,\theta_2,\cdots,\theta_m)=0,\\ \frac{\partial}{\partial\theta_2}\ln L(\theta_1,\theta_2,\cdots,\theta_m)=0,\\ \vdots\\ \frac{\partial}{\partial\theta_m}\ln L(\theta_1,\theta_2,\cdots,\theta_m)=0.\end{cases}\tag{8.11}$$

【Example 8.13】 Let $x_1,x_2,\cdots,x_n$ be the values of a random sample from a normal population with mean μ and variance σ^2. Find the maximum likelihood estimator of the parameters μ and σ^2.

Solution: The probability density function of normal distribution with mean μ and variance σ^2 is

$$f(x;\mu,\sigma^2)=\frac{1}{\sqrt{2\pi}\sigma}e^{-\frac{(x-\mu)^2}{2\sigma^2}},-\infty<x<+\infty.$$

The likelihood function for μ and σ^2 is

$$L(\mu,\sigma^2)=\prod_{i=1}^{n}f(x_i;\mu,\sigma^2)=\prod_{i=1}^{n}\frac{1}{\sqrt{2\pi}\sigma}e^{-\frac{(x_i-\mu)^2}{2\sigma^2}}=\left(\frac{1}{\sqrt{2\pi}\sigma}\right)^n e^{-\frac{1}{2\sigma^2}\sum_{i=1}^{n}(x_i-\mu)^2},$$

$$\ln L(\mu,\sigma^2)=-n(\ln\sigma+\ln\sqrt{2\pi})-\frac{1}{2\sigma^2}\sum_{i=1}^{n}(x_i-\mu)^2.$$

Set the partial differentiations of $\ln L(\mu,\sigma^2)$ with respect to μ and σ^2 equal to 0 to get

$$\begin{cases}\frac{\partial}{\partial\mu}\ln L(\mu,\sigma^2)=\frac{1}{\sigma^2}\sum_{i=1}^{n}(x_i-\mu)=0,\\ \frac{\partial}{\partial\sigma^2}\ln L(\mu,\sigma^2)=-\frac{n}{2\sigma^2}+\frac{1}{2\sigma^4}\sum_{i=1}^{n}(x_i-\mu)^2=0.\end{cases}$$

Solve the above system of equations to get

$$\begin{cases}\hat{\mu}=\frac{1}{n}\sum_{i=1}^{n}x_i=\bar{x},\\ \hat{\sigma}^2=\frac{1}{n}\sum_{i=1}^{n}(x_i-\bar{x})^2=b_2.\end{cases}$$

Therefore, the maximum likelihood estimators for parameters μ and σ^2 is

$$\hat{\mu}=\bar{X},\ \hat{\sigma}^2=\frac{1}{n}\sum_{i=1}^{n}(X_i-\bar{X})^2=B_2.$$

8.3 Interval Estimation[1]

Point estimation is an important and common method in estimation. Nevertheless, a point estimate is a single number, which provides no information about the precision. For example, we use the point estimate $\bar{x}$ for the true average breaking strength μ of paper towels of a certain brand, and suppose that $\bar{x}=9421.4$. Because of sampling variability, it is never the case that $\bar{x}=\mu$, and the point estimate $\bar{x}=9421.4$ says nothing about how close it might be to μ.

There are many situations in which it is preferable to determine an interval within which we would expect to find the value of the parameter. Such an interval is called an interval estimate or **confidence interval**[2].

8.3.1 Basic Concepts of Confidence Intervals

The basic concepts and properties of confidence intervals (CIs) are most easily introduced by first focusing on a simple, albeit somewhat unrealistic, problem situation. Suppose that the parameter of interest is a normal population mean μ with variance σ^2 known. By Corollary 7.1, the sampling distribution of $\bar{X}$ is also a normal distribution with mean μ and variance $\frac{\sigma^2}{n}$. Thus $Z=\frac{\bar{X}-\mu}{\sigma/\sqrt{n}}$ has a standard normal distribution.

Because the area under the standard normal curve between -1.96 and 1.96 is 0.95, that is

$$P\left\{-1.96<\frac{\bar{X}-\mu}{\sigma/\sqrt{n}}<1.96\right\}=0.95. \tag{8.12}$$

The above equation could be rewritten as

$$P\left\{-1.96\cdot\frac{\sigma}{\sqrt{n}}<\bar{X}-\mu<1.96\cdot\frac{\sigma}{\sqrt{n}}\right\}=0.95. \tag{8.13}$$

Which leads to the result

$$P\left\{\bar{X}-1.96\cdot\frac{\sigma}{\sqrt{n}}<\mu<\bar{X}+1.96\cdot\frac{\sigma}{\sqrt{n}}\right\}=0.95. \tag{8.14}$$

Think of a random interval having left endpoint $\bar{X}-1.96\cdot\frac{\sigma}{\sqrt{n}}$ and right endpoint $\bar{X}+1.96\cdot\frac{\sigma}{\sqrt{n}}$, this becomes $\left(\bar{X}-1.96\cdot\frac{\sigma}{\sqrt{n}},\bar{X}+1.96\cdot\frac{\sigma}{\sqrt{n}}\right)$ in interval notation.

---

[1] interval estimation：区间估计

[2] confidence interval：置信区间

The interval $\left(\overline{X}-1.96\cdot\frac{\sigma}{\sqrt{n}},\overline{X}+1.96\cdot\frac{\sigma}{\sqrt{n}}\right)$ is random because the two endpoints of the interval involve a random variable $\overline{X}$. The interval's width is a fixed number $2\times1.96\cdot\frac{\sigma}{\sqrt{n}}$. Now (8.14) can be paraphrased as "the probability that the random interval $\left(\overline{X}-1.96\cdot\frac{\sigma}{\sqrt{n}},\overline{X}+1.96\cdot\frac{\sigma}{\sqrt{n}}\right)$ includes or covers the true value of μ is 0.95". Before any sample observation is gathered, it is quite likely that μ will lie inside this interval. The random interval is an interval estimate for μ, which is called **confidence interval.** The probability 0.95 is called the **degree of confidence.**

Definition 8.9 Suppose that θ is a parameter of a population $X, X_1, X_2, \cdots, X_n$ is a random sample from this population. $\hat{\theta}_L=\hat{\theta}_L(X_1, X_2, \cdots, X_n)$ and $\hat{\theta}_U=\hat{\theta}_U(X_1, X_2, \cdots, X_n)$ are two statistics. If for a given $\alpha(0<\alpha<1)$, we have

$$P\{\hat{\theta}_L<\theta<\hat{\theta}_U\}=1-\alpha. \tag{8.15}$$

The random interval $[\hat{\theta}_L, \hat{\theta}_U]$ is called a $(1-\alpha)100\%$ **confidence interval** for θ.

Moreover, $\hat{\theta}_L$ and $\hat{\theta}_U$ are called **lower and upper confidence limits**[1], $1-\alpha$ is called the **degree of confidence**[2], and the α is called **confidence level**[3].

What does it mean to say you are 95% confident that the true value of the parameter θ is within a given interval? If you construct 20 confidence intervals for θ each using different sample values. You might expect that 95% of them, or 19 of 20, will contain the true value of θ. If you construct 100 intervals, you will expect about 95 of them to perform as planned. You cannot be absolutely sure that any particular interval contains the true value of θ. You will never know whether your particular interval is the one that worked or missed. Your confidence in the estimated interval follows from the fact that when repeated intervals are calculated, 95% of them will contain the true value of θ.

A good confidence interval has two desirable characteristics: Firstly, it is as narrow as possible. The narrower the interval, the more exactly you have located the estimated parameter. Secondly, it has a large confidence degree, near 1. The large the degree of confidence, the more likely it is that the interval will contain the estimated parameter.

The confidence interval is not only one with a specific confidence degree. For

---

[1] lower and upper confidence limits: 置信下限和上限

[2] degree of confidence: 置信度

[3] confidence level: 置信水平

example, in the above example, $1-\alpha=0.95$,

$$P\left\{-1.75<\frac{\overline{X}-\mu}{\sigma/\sqrt{n}}<2.33\right\}=0.95. \tag{8.16}$$

Hence,

$$P\left\{\overline{X}-1.75\cdot\frac{\sigma}{\sqrt{n}}<\mu<\overline{X}+2.33\cdot\frac{\sigma}{\sqrt{n}}\right\}=0.95. \tag{8.17}$$

So, we can get another 95% confidence interval $\left(\overline{X}-1.75\cdot\frac{\sigma}{\sqrt{n}},\overline{X}+2.33\cdot\frac{\sigma}{\sqrt{n}}\right)$ for μ. The width of this interval is $4.08\cdot\frac{\sigma}{\sqrt{n}}$, which is longer than the width of the above interval $3.92\cdot\frac{\sigma}{\sqrt{n}}$. Therefore, we prefer the confidence interval $\left(\overline{X}-1.96\cdot\frac{\sigma}{\sqrt{n}},\overline{X}+1.96\cdot\frac{\sigma}{\sqrt{n}}\right)$ than $\left(\overline{X}-1.75\cdot\frac{\sigma}{\sqrt{n}},\overline{X}+2.33\cdot\frac{\sigma}{\sqrt{n}}\right)$.

How to derive a confidence interval? Let $X_1, X_2, \cdots, X_n$ be the sample on which the CI for a parameter θ is to be based.

Step 1 We should find a random variable denoted as $h(X_1,X_2,\cdots,X_n;\theta)$, which is a function of both $X_1,X_2,\cdots,X_n$ and θ, and the distribution of it does not depend on θ or on any other unknown parameters.

Step 2 For any α between 0 and 1, constants a and b can be found to satisfy

$$P\{a<h(X_1,X_2,\cdots,X_n)<b\}=1-\alpha. \tag{8.18}$$

Suppose that the inequalities in formula (8.18) can be manipulated to isolate θ, giving the equivalent probability statement

$$P\{\hat{\theta}_L(X_1,X_2,\cdots,X_n)<\theta<\hat{\theta}_U(X_1,X_2,\cdots,X_n)\}=1-\alpha.$$

Then $\hat{\theta}_L(X_1,X_2,\cdots,X_n)$ and $\hat{\theta}_U(X_1,X_2,\cdots,X_n)$ are the lower and upper confidence limits for a $100(1-\alpha)\%$ CI.

The function $h(X_1,X_2,\cdots,X_n;\theta)$ is usually considered from a point estimator for parameter θ. The CI for the parameters of population normal can be derived as above process.

8.3.2 Confidence Intervals for Parameters of a Normal Population

1. Confidence intervals for the mean μ

In fact, any desired degree of confidence can be achieved by replacing 1.96 in formula (8.12) with an appropriate standard normal critical value. Writing $z_{\alpha/2}$ for the z-value above which we find an area of $\frac{\alpha}{2}$ under the normal curve. As Figure

8. 3 shows, a probability of $1-\alpha$ is achieved by using $z_{\alpha/2}$ in place of 1. 96,

$$P\left\{-z_{\alpha/2}<\frac{\overline{X}-\mu}{\sigma/\sqrt{n}}<z_{\alpha/2}\right\}=1-\alpha. \tag{8.19}$$

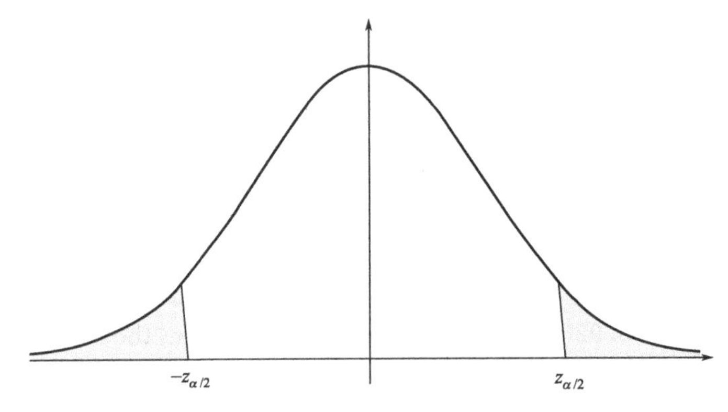

Figure 8. 3 $P\{-z_{\alpha/2}<Z<z_{\alpha/2}\}=1-\alpha$

Hence

$$P\left\{\overline{X}-z_{\alpha/2}\cdot\frac{\sigma}{\sqrt{n}}<\mu<\overline{X}+z_{\alpha/2}\cdot\frac{\sigma}{\sqrt{n}}\right\}=1-\alpha. \tag{8.20}$$

We summarize the discussions above as below.

Confidence Interval on μ, σ^2 known

A $100(1-\alpha)\%$ confidence interval for the mean μ of a normal population with a known variance σ^2 is given by

$$\left(\overline{X}-z_{\alpha/2}\cdot\frac{\sigma}{\sqrt{n}},\overline{X}+z_{\alpha/2}\cdot\frac{\sigma}{\sqrt{n}}\right) \tag{8.21}$$

or, equivalently, by $\left(\overline{X}\pm z_{\alpha/2}\cdot\frac{\sigma}{\sqrt{n}}\right)$.

【Example 8. 14】 The average zinc concentration recovered from a sample of measurements taken in 36 different locations in a river is found to be 2. 6 grams per milliliter. Find the 95% and 99% confidence intervals for the mean zinc concentration in the river. Assume that the population is normal with variance 0. 09.

Solution: In this sample, $n=36$, $\overline{x}=2.6$ and $\sigma=\sqrt{0.09}=0.3$. From the table of the standard normal distribution, we can find $z_{0.025}=1.96$. Hence, the 95% confidence interval for the mean is $\left(2.6-1.96\cdot\frac{0.3}{\sqrt{36}},2.6+1.96\cdot\frac{0.3}{\sqrt{36}}\right)$, which reduces to (2. 50, 2. 70).

To find a 99% confidence interval, we find $z_{0.005}=2.575$, and the 99% confidence interval is $\left(2.6-2.575\cdot\frac{0.3}{\sqrt{36}},2.6+2.575\cdot\frac{0.3}{\sqrt{36}}\right)$, or simply (2. 47, 2. 73).

From above example we see that a longer interval is required to estimate μ with a higher degree of confidence.

Frequently, the population variance σ^2 is often unknown. Recalling the result in Chapter 7 that if we have a random sample from a normal distribution, then the random variable $T=\dfrac{\overline{X}-\mu}{S/\sqrt{n}}$ has a Student t-distribution with $n-1$ degrees of freedom. T can be used to construct a confidence interval on μ. The procure is the same as that with σ^2 known except that σ is replaced by S and the standard normal distribution is replaced by the t-distribution. Referring to Figure 8.4, we can see that

$$P\{-t_{\alpha/2}(n-1)<T<t_{\alpha/2}(n-1)\}=1-\alpha. \tag{8.22}$$

Where $t_{\alpha/2}(n-1)$ is the t-value with n−1 degrees of freedom, above which we find an area of $\frac{\alpha}{2}$. Substituting for T, we write

$$P\{-t_{\alpha/2}(n-1)<\frac{\overline{X}-\mu}{S/\sqrt{n}}<t_{\alpha/2}(n-1)\}=1-\alpha. \tag{8.23}$$

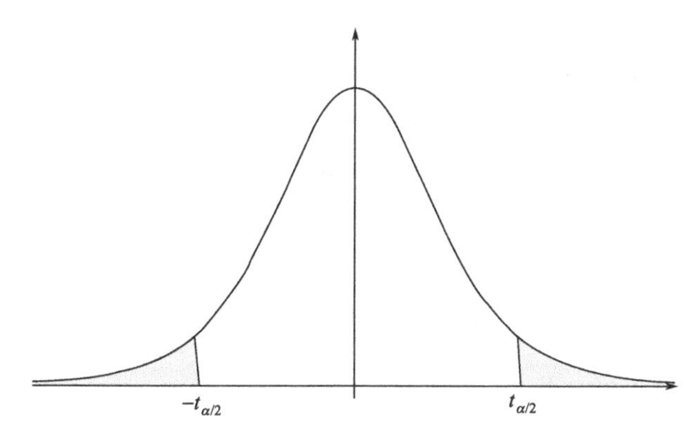

Figure 8.4 $P\{-t_{\alpha/2}(n-1)<T<t_{\alpha/2}(n-1)\}=1-\alpha$

Hence

$$P\{\overline{X}-t_{\alpha/2}(n-1)\cdot\frac{S}{\sqrt{n}}<\mu<\overline{X}+t_{\alpha/2}(n-1)\cdot\frac{S}{\sqrt{n}}\}=1-\alpha. \tag{8.24}$$

Thus, the confidence interval on μ with σ^2 unknown is as follows.

Confidence Interval on μ, σ^2 unknown

A $100(1-\alpha)\%$ confidence interval for the mean μ of a normal population with unknown variance σ^2 is given by

$$\left(\overline{X}-t_{\alpha/2}(n-1)\cdot\frac{S}{\sqrt{n}},\overline{X}+t_{\alpha/2}(n-1)\cdot\frac{S}{\sqrt{n}}\right) \tag{8.25}$$

or, equivalently, by $\left(\overline{X}\pm t_{\alpha/2}(n-1)\cdot\frac{S}{\sqrt{n}}\right)$.

【Example 8.15】 The contents of seven similar containers of sulfuric acid are 9.8, 10.2, 10.4, 9.8, 10.0, 10.2 and 9.6 liters. Find a 95% confidence interval for the mean contents of all such containers, assuming an approximately normal distribution.

Solution: The sample mean and standard deviation for the given data are

$$\bar{x}=10.0 \text{ and } s=0.283.$$

Using appendix Table 4, we can find $t_{0.025}(6)=2.447$. Hence, the 95% confidence interval for μ is

$$\left(10.0-2.447\cdot\frac{0.283}{\sqrt{7}},10.0+2.447\cdot\frac{0.283}{\sqrt{7}}\right)$$

which reduces to (9.74, 10.26).

You may find that confidence interval on μ with σ^2 unknown is more general than that with σ^2 known and wonder why we do not use formula (8.25) for all of the problems even in the cases when the population variance σ^2 is known. It is because in those cases, the confidence interval given by formula (8.21) is shorter than the one given by (8.25), which means "better" in estimation.

2. Confidence intervals for the variance σ^2

For a random sample from a normal population with variance σ^2, the sample variance S^2 is an unbiased point estimator of σ^2. An interval estimator of σ^2 can be established by using the statistic $\chi^2=\frac{(n-1)S^2}{\sigma^2}$. According to Theorem 7.8, the statistic $\chi^2=\frac{(n-1)S^2}{\sigma^2}$ has a chi-squared distribution with $n-1$ degrees of freedom. Referring to Figure 8.5, we may write

$$P\{\chi^2_{1-\alpha/2}(n-1)<\chi^2<\chi^2_{\alpha/2}(n-1)\}=1-\alpha \tag{8.26}$$

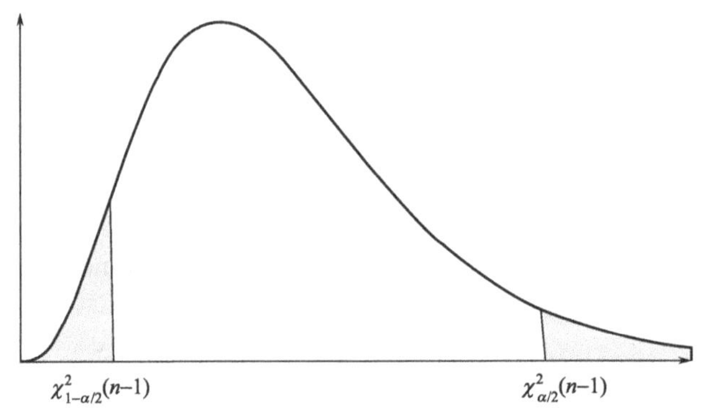

Figure 8.5 $P\{\chi^2_{1-\alpha/2}(n-1)<\chi^2<\chi^2_{\alpha/2}(n-1)\}=1-\alpha$

where $\chi^2_{1-\alpha/2}(n-1)$ and $\chi^2_{\alpha/2}(n-1)$ are values of the chi-squared distribution with $n-1$ degrees of freedom above which we find areas of $1-\frac{\alpha}{2}$ and $\frac{\alpha}{2}$.

Substituting for χ^2, we write

$$P\{\chi^2_{1-\alpha/2}(n-1)<\frac{(n-1)S^2}{\sigma^2}<\chi^2_{\alpha/2}(n-1)\}=1-\alpha. \tag{8.27}$$

Dividing each term in the inequality by $(n-1)S^2$ and then inverting each term, we obtain

$$P\{\frac{(n-1)S^2}{\chi^2_{\alpha/2}(n-1)}<\sigma^2<\frac{(n-1)S^2}{\chi^2_{1-\alpha/2}(n-1)}\}=1-\alpha. \tag{8.28}$$

Thus, the confidence interval on variance σ^2 of a normal distribution is as follows.

Confidence Interval for σ^2

A $100(1-\alpha)\%$ confidence interval for σ^2 of a normal population is given by

$$\left(\frac{(n-1)S^2}{\chi^2_{\alpha/2}(n-1)},\frac{(n-1)S^2}{\chi^2_{1-\alpha/2}(n-1)}\right). \tag{8.29}$$

A $100(1-\alpha)\%$ confidence interval for σ is obtained by taking the square root of each endpoint of the interval for σ^2.

【Example 8.16】 A study about the lifetime of a kind of light bulb has a random sample of 17 bulbs with the sample standard deviation of 3.2 hours. Construct a 98% confidence interval for the population variance σ^2, assuming a normal population.

Solution: Here, $n=17$ and $S=3.2$. To obtain a 98% confidence interval, we choose $\alpha=0.02$. Then, using appendix Table 3 with 16 degrees of freedom, we can find $\chi^2_{0.01}(16)=32.00$ and $\chi^2_{0.99}(16)=5.812$. Therefore, the 98% confidence interval for σ^2 is

$$\left(\frac{(17-1)3.2^2}{32},\frac{(17-1)3.2^2}{5.812}\right) \text{ or simply } (3.2,\ 28.19).$$

8.3.3 Confidence Intervals for the Difference of the Sample Means $\mu_1-\mu_2$

If we have two normal populations $N(\mu_1,\sigma_1^2)$ and $N(\mu_2,\sigma_2^2)$. Suppose that $\overline{X}_1$ is the sample mean of a random sample of size n_1 from the normal population $N(\mu_1,\sigma_1^2)$, $\overline{X}_2$ is the sample mean of a random sample of size n_2 from the normal population $N(\mu_2,\sigma_2^2)$, and these two samples are independent. According to Corollary 7.1, $\overline{X}_1$ and $\overline{X}_2$ have normal populations $N\left(\mu_1,\frac{\sigma_1^2}{n_1}\right)$ and $N\left(\mu_2,\frac{\sigma_2^2}{n_2}\right)$, respectively. Therefore,

$$Z=\frac{(\overline{X}_1-\overline{X}_2)-(\mu_1-\mu_2)}{\sqrt{\sigma_1^2/n_1+\sigma_2^2/n_2}}\sim N(0,1).$$

Referring once again to Figure 8.3, we see that

$$P\left\{-z_{\alpha/2}<\frac{(\overline{X}_1-\overline{X}_2)-(\mu_1-\mu_2)}{\sqrt{\sigma_1^2/n_1+\sigma_2^2/n_2}}<z_{\alpha/2}\right\}=1-\alpha. \tag{8.30}$$

Hence

$$P\left\{(\overline{X}_1-\overline{X}_2)-z_{\alpha/2}\sqrt{\frac{\sigma_1^2}{n_1}+\frac{\sigma_2^2}{n_2}}<\mu_1-\mu_2<(\overline{X}_1-\overline{X}_2)+z_{\alpha/2}\sqrt{\frac{\sigma_1^2}{n_1}+\frac{\sigma_2^2}{n_2}}\right\}=1-\alpha \tag{8.31}$$

which leads to the following confidence interval for $\mu_1-\mu_2$.

Confidence Interval for $\mu_1-\mu_2$, σ_1^2 and σ_2^2 Known

A $100(1-\alpha)\%$ confidence interval for $\mu_1-\mu_2$ is given by

$$\left((\overline{X}_1-\overline{X}_2)-z_{\alpha/2}\sqrt{\frac{\sigma_1^2}{n_1}+\frac{\sigma_2^2}{n_2}},(\overline{X}_1-\overline{X}_2)+z_{\alpha/2}\sqrt{\frac{\sigma_1^2}{n_1}+\frac{\sigma_2^2}{n_2}}\right) \tag{8.32}$$

or, equivalently, by $\left((\overline{X}_1-\overline{X}_2)\pm z_{\alpha/2}\sqrt{\frac{\sigma_1^2}{n_1}+\frac{\sigma_2^2}{n_2}}\right)$.

【Example 8.17】 We want to compare the consumption of two different brand toner cartridges. Given that a random sample of 20 toner cartridges of brand A has the average 1520 sheets per cartridge and another random sample 30 toner cartridges of brand B has the average 1445 sheets per cartridge. If the population standard deviations are given as $\sigma_A=12$ and $\sigma_B=15$, find a 95% confidence interval for the difference of the population means $\mu_1-\mu_2$.

Solution: For $\alpha=0.05$, we know that $z_{0.025}=1.96$. Thus, a 95% confidence interval for the difference of the population means $\mu_1-\mu_2$ is given by

$$\left((1520-1445)-1.96\sqrt{\frac{12^2}{20}+\frac{15^2}{30}},(1520-1445)+1.96\sqrt{\frac{12^2}{20}+\frac{15^2}{30}}\right)$$

which reduces to (47.49, 52.51).

Consider the case where σ_1^2 and σ_2^2 are unknown. If $\sigma_1^2=\sigma_2^2=\sigma^2$, according to Corollary 7.4, $T=\frac{(\overline{X}_1-\overline{X}_2)-(\mu_1-\mu_2)}{S_W\sqrt{1/n_1+1/n_2}}\sim t(n_1+n_2-2)$. Referring once again to Figure 8.4, we have

$$P\left\{-t_{\alpha/2}(n_1+n_2-2)<\frac{(\overline{X}_1-\overline{X}_2)-(\mu_1-\mu_2)}{S_W\sqrt{1/n_1+1/n_2}}<t_{\alpha/2}(n_1+n_2-2)\right\}=1-\alpha \tag{8.33}$$

where $S_W^2=\dfrac{(n_1-1)S_1^2+(n_2-1)S_2^2}{n_1+n_2-2}$, $t_{\alpha/2}(n_1+n_2-2)$ is the t-value with n_1+n_2-2 degrees of freedom, above which we find an area of $\dfrac{\alpha}{2}$.

Hence,

$$P\{(\overline{X}_1-\overline{X}_2)-t_{\alpha/2}(n_1+n_2-2)S_W\sqrt{1/n_1+1/n_2}<\mu_1-\mu_2<(\overline{X}_1-\overline{X}_2)+t_{\alpha/2}(n_1+n_2-2)S_W\sqrt{1/n_1+1/n_2}\}=1-\alpha.$$

Confidence Interval for $\mu_1-\mu_2$, $\sigma_1^2=\sigma_2^2$ but Both Unknown

Suppose two normal populations with unknown but equal variances, a $100(1-\alpha)\%$ confidence interval for $\mu_1-\mu_2$ is given by

$$\left((\overline{X}_1-\overline{X}_2)-t_{\alpha/2}(n_1+n_2-2)S_W\sqrt{\frac{1}{n_1}+\frac{1}{n_2}},\right.$$
$$\left.(\overline{X}_1-\overline{X}_2)+t_{\alpha/2}(n_1+n_2-2)S_W\sqrt{\frac{1}{n_1}+\frac{1}{n_2}}\right) \tag{8.34}$$

or, equivalently, by $\left((\overline{X}_1-\overline{X}_2)\pm t_{\alpha/2}(n_1+n_2-2)S_W\sqrt{\dfrac{1}{n_1}+\dfrac{1}{n_2}}\right)$.

【Example 8.18】 Two independent sampling stations were chosen for the study of evaluating the effectiveness of a numerical species diversity index to indicate aquatic degradation due to acid mine drainage. One located downstream from the acid mine discharge point and the other located upstream. For 12 monthly samples collected at the downstream station, the species diversity index had a mean value $\bar{x}_1=3.11$ a standard deviation $s_1=0.771$, while 10 monthly samples collected at the upstream station had a mean index value $\bar{x}_2=2.04$ and a standard deviation $s_2=0.448$. Find a 90% confidence interval for the difference between the population means for the two locations, assuming that the populations are normally distributed with equal variances.

Solution: Let μ_1 and μ_2 represent the population means, respectively, for the species diversity indices at the downstream and upstream stations.

$\bar{x}_1-\bar{x}_2=3.11-2.04=1.07$,

$$S_W^2=\frac{(n_1-1)s_1^2+(n_2-1)s_2^2}{n_1+n_2-2}=\frac{(12-1)0.771^2+(10-1)0.448^2}{12+10-2}=0.417.$$

Taking the square root, we obtain $s_W=0.646$. Using $\alpha=0.1$, we find in appendix Table 4 that $t_{0.05}(n_1+n_2-2)=t_{0.05}(20)=1.725$. Therefore, the 90% confidence interval for $\mu_1-\mu_2$ is

$$\left(1.07-1.725\cdot 0.646\cdot\sqrt{\frac{1}{12}+\frac{1}{10}},1.07+1.725\cdot 0.646\cdot\sqrt{\frac{1}{12}+\frac{1}{10}}\right)$$

which simplifies to (0.593, 1.547).

Let us now consider the problem of finding an interval estimate of $\mu_1 - \mu_2$ when the unknown population variances are not likely to be equal. In this case, the statistic often used is

$$T' = \frac{(\overline{X}_1 - \overline{X}_2) - (\mu_1 - \mu_2)}{\sqrt{S_1^2/n_1 + S_2^2/n_2}}$$

which has approximately a t-distribution with v degrees of freedom, where

$$v = \frac{(s_1^2/n_1 + s_2^2/n_2)^2}{(s_1^2/n_1)^2/(n_1-1) + (s_2^2/n_2)^2/(n_2-1)}.$$

Since v is seldom an integer, we round it down to the nearest whole number. Referring to Figure 8.4, we have

$$P\{-t_{\alpha/2}(v) < T' < t_{\alpha/2}(v)\} = 1-\alpha. \tag{8.35}$$

Substituting for T' in the inequality and following the same steps as before, we state the final result.

Confidence Interval for $\mu_1 - \mu_2$, $\sigma_1^2 \neq \sigma_2^2$ and Both Unknown

Suppose two normal populations with unknown and unequal variances, a 100 $(1-\alpha)\%$ confidence interval for $\mu_1 - \mu_2$ is given by

$$\left((\overline{X}_1 - \overline{X}_2) - t_{\alpha/2}(v)\sqrt{\frac{S_1^2}{n_1} + \frac{S_2^2}{n_2}}, (\overline{X}_1 - \overline{X}_2) + t_{\alpha/2}(v)\sqrt{\frac{S_1^2}{n_1} + \frac{S_2^2}{n_2}}\right) \tag{8.36}$$

where $v = \dfrac{(s_1^2/n_1 + s_2^2/n_2)^2}{(s_1^2/n_1)^2/(n_1-1) + (s_2^2/n_2)^2/(n_2-1)}$.

【Example 8.19】 A study was conducted to estimate the difference in the amounts of the chemical organophosphorus measured at two different stations on the James River. Organophosphorus was measured in milligrams per liter. 15 samples were collected from station 1, and 12 samples were obtained from station 2. The 15 samples from station 1 had an average organophosphorus content of 3.84 milligrams per liter and a standard deviation of 3.07 milligrams per liter, while the 12 samples from station 2 had an average content of 1.49 milligrams per liter and a standard deviation of 0.80 milligram per liter. Find a 95% confidence interval for the difference in the true average organophosphorus contents at these two stations, assuming that the observations came from normal populations with different variances.

Solution: For station 1, we have $n_1 = 15$, $\overline{x}_1 = 3.84$ and $s_1 = 3.07$. For station 2, $n_2 = 12$, $\overline{x}_2 = 1.49$ and $s_2 = 0.80$.

We wish to find 95% confidence interval for $\mu_1 - \mu_2$. Since the population variances are assumed to be unequal, we can only find and approximate 95% confidence interval based on the t-distribution with v degrees of freedom, where

$$v=\frac{(3.07^2/15+0.80^2/12)^2}{(3.07^2/15)^2/14+(0.80^2/12)^2/11}=16.3\approx 16.$$

Using $\alpha=0.05$, we find in appendix Table 4 that $t_{0.025}(16)=2.120$. Therefore, the 95% confidence interval for $\mu_1-\mu_2$ is

$$\left[(3.84-1.49)-2.120\cdot\sqrt{\frac{3.07^2}{15}+\frac{0.80^2}{12}},(3.84-1.49)+2.120\cdot\sqrt{\frac{3.07^2}{15}+\frac{0.80^2}{12}}\right]$$

which simplifies to (0.60, 4.10).

8.4 Confidence Interval for a Population Proportion p

Many research experiments or sample surveys need to estimate the proportion of people or objects in a large group that possess a certain characteristic. Here are some examples:

- The proportion of sales that can be expected in a large number of customer contacts.
- The proportion of seeds that germinate.
- The proportion of "likely" voters who plan to vote for a particular political candidate.

All of the above examples are the binomial experiment, and the parameter to be estimated is the binomial proportion p.

The best point estimator for the population proportion p is given by $\hat{p}=\frac{X}{n}$, where X represents the number of successes in n trials. Therefore, the sample proportion $\hat{p}=\frac{x}{n}$ will be used as the point estimate of the parameter p.

According to Central Limit Theorem, for n sufficiently large, $\hat{p}$ is approximately normally distributed with mean $E(\hat{p})=E\left(\frac{X}{n}\right)=\frac{np}{n}=p$ and variance

$$Var(\hat{p})=Var\left(\frac{X}{n}\right)=\frac{np(1-p)}{n^2}=\frac{p(1-p)}{n}.$$

Therefore, we can assert that

$$P\left\{-z_{\alpha/2}<\frac{\hat{p}-p}{\sqrt{p(1-p)/n}}<z_{\alpha/2}\right\}=1-\alpha. \tag{8.37}$$

Following the same steps as before, we can obtain the result as follows.

Confidence Interval for a Population Proportion p

An approximate $100(1-\alpha)\%$ confidence interval for a population proportion p is given by

$$\hat{p} \pm z_{\alpha/2}\sqrt{\frac{p(1-p)}{n}}. \tag{8.38}$$

since p is unknown, it is estimated using the best point estimator $\hat{p}$.

【Example 8.20】 A random sample of 985 "likely" voters—those who are likely to vote in the upcoming election—were polled during a Phone-Athon conducted by the Republican Party in US. 592 indicated that they intended to vote for the Republican candidate in the upcoming election. Construct a 90% confidence interval for p, the proportion of likely voters in the population who intend to vote for the Republican candidate. Based on this information, can you conclude that the candidate will win the election?

Solution: The point estimate for p

$$\hat{p}=\frac{x}{n}=\frac{592}{985}=0.601.$$

And the estimated standard error is

$$\sqrt{\frac{\hat{p}(1-\hat{p})}{n}}=\sqrt{\frac{0.601\times(1-0.601)}{985}}=0.016.$$

Using $\alpha=0.1$, we find in appendix Table 2 that $z_{0.05}=1.645$. Therefore, the 90% confidence interval for p is thus

$$(0.601-1.645\times0.016, 0.601+1.645\times0.016)$$

which simplifies to (0.575, 0.627).

The percentage of likely voters who intend to vote for the Republican candidate is between 57.5% and 62.7%. Assuming that she needs more than 50% of the vote to win, and since both the upper and lower confidence limits exceed this minimum value, we can say with 90% confidence that the candidate will win.

Exercises

1. Suppose a random sample $X_1, X_2, \cdots, X_n$ from a population with mean μ. For which constants a and b is $T=a(X_1+X_2+\cdots+X_n)+b$ an unbiased estimator for μ?
2. Let $X_1, X_2, \cdots, X_n$ be a random sample from a population with mean μ. Sup-

pose that the population has a finite variance σ^2, prove that $\frac{1}{n}\sum_{i=1}^{n}(X_i-\mu)^2$ is an unbiased estimator of σ^2.

3. $X_1, X_2, \cdots, X_4$ is a random sample from a normal population $N(\mu, \sigma^2)$. Three estimators are defined as

$$T_1=\frac{1}{4}X_1+\frac{1}{4}X_2+\frac{1}{4}X_3+\frac{1}{4}X_4,$$

$$T_2=\frac{1}{2}X_1+\frac{1}{8}X_2+\frac{1}{8}X_3+\frac{1}{4}X_4,$$

$$T_3=\frac{1}{3}X_1+\frac{1}{4}X_2+\frac{1}{6}X_3+\frac{1}{4}X_4.$$

(1) Prove that T_1, T_2, T_3 are all unbiased estimator of the μ;

(2) Considering estimators T_1, T_2, T_3, find which one is the best estimator.

4. Let $\overline{X}$ be the sample mean of a random sample of size n from an exponential population with parameter λ. Prove that $\overline{X}$ is a minimum variance unbiased estimator of the parameter λ.

5. Let $X_1, X_2, \cdots, X_n$ be a random sample from a population with mean μ. We consider the estimator $\hat{\mu}_n \sim N\left(\mu-\frac{1}{n}, \frac{1}{n}\right)$

(1) Show that $\hat{\mu}_n$ is a biased estimator for μ.

(2) Show that $\hat{\mu}_n$ is a consistent estimator for μ.

6. Let $X_1, X_2, \cdots, X_n$ be a random sample from a uniform population X on $[\alpha, \beta]$, with $\alpha=1$, find an estimator of the parameter β by the method of moments.

7. A random sample from a population X has the observed values $x_1=1, x_2=2$, $x_3=1$. The probability mass distribution of X is

X	1	2	3
P	θ^2	$2\theta(1-\theta)$	$(1-\theta)^2$

with $\theta(0<\theta<1)$ unknown. Find an estimate of the parameter θ by the method of moments.

8. Given a random sample $X_1, X_2, \cdots, X_n$ from a population whose probability density function is defined as

$$f(x)=\begin{cases}\sqrt{\theta}x^{\sqrt{\theta}-1}, & 0\leqslant x\leqslant 1\\ 0, & \text{elsewhere}\end{cases}$$ with θ unknown and $\theta>0$.

Find an estimator of parameter θ by the method of moments.

9. The lifetimes (number of weeks to breakdown, given continual use) of widgets

have an exponential distribution with unknown parameter λ. The factory producing the widgets measures the lifetimes of a random sample of 10 widgets as

5.010, 15.300, 3.816, 1.476, 3.378, 5.622, 1.350, 12.816, 2.100, 7.368.

Find the maximum likelihood estimate of parameter λ.

10. If $x_1, x_2, \cdots, x_n$ are observed values of a random sample from a population with probability mass function given by

$$P\{X=x_k\}=\binom{x_k-1}{r-1}p^r(1-p)^{x_k-r}, x_k=r, r+1, \cdots$$

with parameter r known and p unknown. Find the maximum likelihood estimate of parameter p.

11. If $X_1, X_2, \cdots, X_n$ is a random sample from a population with probability density function given by

$$f(x)=\begin{cases}\dfrac{1}{\theta}e^{\frac{\delta-x}{\theta}}, & \text{for } x>\delta,\\ 0, & \text{elsewhere.}\end{cases}$$

Find estimators for δ and θ by method of maximum likelihood estimation.

12. The following data are measurements of plasma citrate concentrations from 9 volunteers (in micromoles per liter):

93, 116, 125, 144, 105, 89, 116, 151, 137.

Assuming that the population of such measurements is normally distributed and has standard deviation 20 micromoles per liter, calculate a 95% confidence interval for the population mean based on these data. If instead a sample of 200 such measurements was available, what would be the length of the corresponding 95% confidence interval?

13. A zoologist is interested in the weights of a population of sparrows. He is able to catch and weigh 90 birds, and their measurements (in gram) $x_1, x_2, \cdots, x_{90}$ can be summarized by $\sum_{i=1}^{90} x_i = 1794$, $\sum_{i=1}^{90} x_i^2 = 35892$. He believes that the weight of the population are normally distributed, and that the measured birds form a random sample from the population. Find a 95% confidence interval for the mean weight of the population.

14. If a random sample of size 9 from normal population has the observation values

12, 15, 14, 21, 18, 16, 10, 9, 11

(1) Find an unbiased point estimate of the population variance σ^2.

(2) Find a 95% confidence interval for the population variance σ^2.

15. We have two independent random samples which are from two normal popula-

tions. The first sample of size 9 is from a population with variance $\sigma_1^2=12$, and it has the sample mean $\bar{x}_1=23$. The second sample of size 15 is from another population also with variance $\sigma_2^2=12$, and the sample mean is $\bar{x}_2=17$. Find a 90% confidence interval for the difference of the population mean $\mu_1-\mu_2$.

16. If in the exercise 15, we don't know the population variances σ_1^2 and σ_2^2, instead we know their sample variance $s_1^2=21$, $s_2^2=14$. Suppose that the two populations have same variance, find a 90% confidence interval for the difference of the population means $\mu_1-\mu_2$.
17. In a political opinion poll, 50 voters are asked whether they are in favor of a particular policy, and 37 of them say that they are. Calculate a 90% confidence interval for the proportion of voters in the whole population who are in favor of the policy. What assumptions are necessary in making this calculation?

Chapter 9 Hypothesis Testing[1]

In practical application, the problem of statistical inference faced by statisticians or engineers is not only the estimation of unknown parameters of a distribution as discussed in Chapter 8, but rather the conclusion to be drawn concerned with two conflicting claims about the parameter. For example, an engineer wants to know whether dry drilling is faster or the same as wet drilling; a medical researcher wishes to decide whether coffee drinking increases the risk of cancer in humans. The process of formulating the possible conclusions one can draw from an experiment and choosing between two alternatives is known as **hypothesis testing.**

9.1 Basic Concepts and Principles of Hypothesis Testing

9.1.1 Hypothesis and Test Statistic[2]

A statistical hypothesis, or just hypothesis, is a claim or assertion either about the value of the parameter concerning a population, or about the form of an entire probability distribution. For example, a hypothesis is the claim $\mu = 68$, where μ is the true average weight of male students in a certain college. Another example is the assertion that $\mu_1 = \mu_2$, where μ_1 and μ_2 is the true average breaking strengths of two different types of twine. In any hypothesis testing problem, there are two conflicting hypotheses under consideration. The objective of hypothesis testing is to decide which of them is correct. Usually, one of the hypotheses is initially favored, which will not be rejected unless sample evidence contradicts it and provides strong support for the alternative claim.

---

[1] hypothesis testing：假设检验

[2] test statistic：检验统计量

Definition 9.1 The **null hypothesis**[1], denoted by H_0, is the claim that is initially assumed to be true. The **alternative hypothesis**[2], denoted by H_1, is the assertion that is contradictory to H_0.

The null hypothesis H_0 will be rejected if sample evidence suggests that H_0 is false. Otherwise, we will continue to believe in the plausibility of the H_0. Hypothesis testing is to use sample data to decide whether the evidence favors H_1 or H_0. The two possible conclusions are then reject H_0 in favor of H_1 or fail to reject H_0.

【Example 9.1】 We wish to know that the average weight of male students in a certain college is different from 68 kilograms. Let μ be the mean of the population. The null hypothesis is

$$H_0: \mu = 68,$$

and the alternative hypothesis is

$$H_1: \mu \neq 68.$$

To illustrate the test of a statistical hypothesis about a population, we take the following example as considered.

【Example 9.2】 When the packing machine which packs salt is working properly, the average weight of salt is 0.5kg, and the standard deviation is 0.015kg. In order to check whether the packing machine is working properly, 9 bags of salt were randomly selected and the weights were (kg) 0.497, 0.506, 0.518, 0.524, 0.498, 0.511, 0.520, 0.515, 0.512, respectively. Can you say the packing machine is working properly (Assume that the standard deviation is relatively stable)?

Let μ, σ be the mean and standard deviation of the population X, the weights of the bagged salt packed in the whole day. Assume the standard deviation of the population of X to be $\sigma = 0.015$, then $X \sim N(\mu, 0.015^2)$ with μ unknown. Consider the null hypothesis that the average weight of the bagged salt is 0.5 against the alternative hypothesis that it is unequal to 0.5. That is,

$$H_0: \mu = \mu_0 = 0.5, \ H_1: \mu \neq \mu_0 = 0.5.$$

From Chapter 8, we know the sample mean $\overline{X}$ is the most efficient estimator of μ. The observed value of $\overline{X}$ reflects the value of μ to some extent. Therefore, we choose $\overline{X}$ as the test statistic of this problem. If the null hypothesis H_0 is true, the observed value of sample mean $\bar{x}$ should not be too far from the population

---

[1] null hypothesis：原假设

[2] alternative hypothesis：备择假设

mean μ_0, i.e. $|\bar{x}-\mu_0|<k_0$, where k_0 is a specified value. If $|\bar{x}-\mu_0|\geqslant k_0$, it is reasonable to doubt H_0, and thus reject H_0 and accept H_1. Since

$$\{|\bar{x}-\mu_0|\geqslant k_0\}=\left\{\left|\frac{\bar{x}-\mu_0}{\sigma/\sqrt{n}}\right|\geqslant\frac{k_0}{\sigma/\sqrt{n}}\right\}=\left\{\left|\frac{\bar{x}-\mu_0}{\sigma/\sqrt{n}}\right|\geqslant k\right\} \tag{9.1}$$

where $k=\dfrac{k_0}{\sigma/\sqrt{n}}$, so we consider $\left|\dfrac{\bar{x}-\mu_0}{\sigma/\sqrt{n}}\right|\geqslant k$ instead of $|\bar{x}-\mu_0|\geqslant k_0$.

In order to decide the value of k, we consider the statistic $Z=\dfrac{\overline{X}-\mu_0}{\sigma/\sqrt{n}}$. If we take $\alpha=0.05$, and let $P\left\{\left|\dfrac{\overline{X}-\mu_0}{\sigma/\sqrt{n}}\right|\geqslant k\right\}=P\{|Z|\geqslant k\}=\alpha$. If the null hypothesis H_0 is true, the statistic $Z=\dfrac{\overline{X}-\mu_0}{\sigma/\sqrt{n}}\sim N(0,1)$, according to the definition of the quantile of the $N(0,1)$ (Figure 9.1), we can get $k=z_{\alpha/2}$. When the observed value of the statistic Z satisfies $|z|\geqslant z_{\alpha/2}$, we would reject the null hypothesis H_0. Otherwise, we would not reject the null hypothesis. Usually, we called the region $|z|\geqslant z_{\alpha/2}$ that we reject the null hypothesis H_0 the **critical region**[1], and the region $|z|<z_{\alpha/2}$ the **acceptance region**[2].

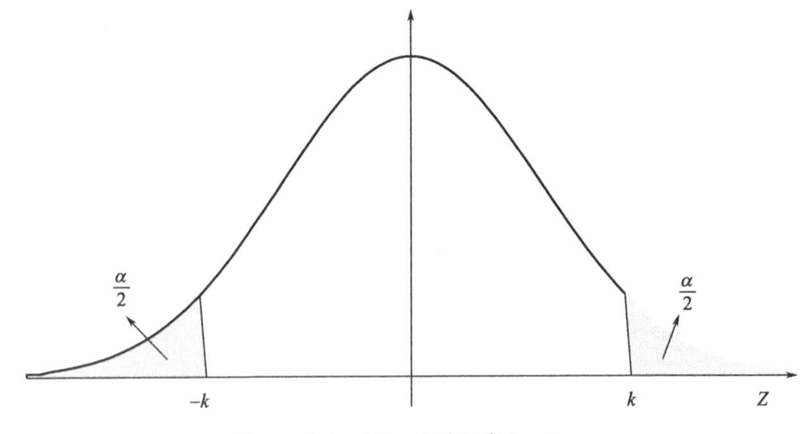

Figure 9.1 $P\{-k\leqslant Z\leqslant k\}=1-\alpha$

The statistic $Z=\dfrac{\overline{X}-\mu_0}{\sigma/\sqrt{n}}$ is called the **test statistic**, the number $k=z_{\alpha/2}$ that

---

[1] critical region: 拒绝域

[2] acceptance region: 接受域

separates the acceptance and critical region is call the **critical value**[1], the value α is called a **significant level.**

In the above example, if we take $\alpha=0.05$, it is easy to find the critical value $z_{\alpha/2}=z_{0.025}=1.96$ from the appendix Table 2. $n=9, \sigma=0.015$, and we can calculate $\bar{x}=0.511$, then the value of the test statistic Z is $|z|=\left|\frac{\bar{x}-\mu_0}{\sigma/\sqrt{n}}\right|=2.2$. Sine $2.2>1.96$, then we have reason to reject H_0, which means we have reason to think that the packing machine is working properly.

9.1.2 Errors in Hypothesis Testing

It should be emphasized that the hypothesis testing can not constitute a mathematical proof of the truth or falsity of the hypothesis. There are two different types of errors that might be made in the course of a statistical hypothesis testing analysis. We would be committing an error by rejecting H_0 in favor of H_1 when H_0 is true, or we fail to reject H_0 when in fact H_0 is false.

Definition 9.2 Rejection of the null hypothesis when it is true is called a **type Ⅰ error**[2]. Nonrejection of the null hypothesis when it is false is called a **type Ⅱ error**[3].

In Example 9.2, the probability of committing a Type Ⅰ error is

$$P\{\text{Type I error}\}=P\{\text{reject } H_0 \mid H_0 \text{ is true}\}=P\{|Z|\geqslant z_{\alpha/2} \mid \mu=0.5\}=\alpha. \tag{9.2}$$

That is, the significance level in the test procedure is the probability of committing a Type Ⅰ error, so the Type Ⅰ error is usually denoted by α. The smaller the value of α, the smaller the chance of committing a Type Ⅰ error.

The probability of committing a Type Ⅱ error is usually denoted by β. That is

$$\beta=P\{\text{Type II error}\}=P\{\text{Do not reject } H_0 \mid H_0 \text{ is false}\}. \tag{9.3}$$

The probability of committing a Type Ⅱ error in Example 9.2 is

$$\begin{aligned}\beta &=P\{\text{Do not reject } H_0 \mid H_0 \text{ is false}\}=P\{|Z|<z_{\alpha/2} \mid \mu\neq 0.5\} \\ &=P\left\{\left|\frac{\bar{X}-0.5}{\sigma/\sqrt{n}}\right|<z_{\alpha/2} \mid \mu\neq 0.5\right\}.\end{aligned} \tag{9.4}$$

Since H_0 is false, then $\bar{X}\sim N(\mu,\sigma^2/n)$ with $\mu\neq 0.5$. Thus

---

[1] critical value：临界值

[2] type Ⅰ error：第一类错误

[3] type Ⅱ error：第二类错误

$$P\left\{\left|\frac{\overline{X}-0.5}{\sigma/\sqrt{n}}\right|<z_{\alpha/2}\,\middle|\,\mu\neq 0.5\right\}=P\left\{\left|\frac{\overline{X}-\mu+\mu-0.5}{\sigma/\sqrt{n}}<z_{\alpha/2}\right|\mu\neq 0.5\right\}$$

$$=P\left\{\frac{0.5-\mu}{\sigma/\sqrt{n}}-z_{\alpha/2}<\frac{\overline{X}-\mu}{\sigma/\sqrt{n}}<\frac{0.5-\mu}{\sigma/\sqrt{n}}+z_{\alpha/2}\right\} \tag{9.5}$$

$$=\Phi\left(\frac{0.5-\mu}{\sigma/\sqrt{n}}+z_{\alpha/2}\right)-\Phi\left(\frac{0.5-\mu}{\sigma/\sqrt{n}}-z_{\alpha/2}\right).$$

It is impossible to compute the Type Ⅱ error β unless we have a specific μ. For $\mu=0.505$,

$$\beta=\Phi\left(\frac{0.5-0.505}{0.015/\sqrt{9}}+1.96\right)-\Phi\left(\frac{0.5-0.505}{0.015/\sqrt{9}}-1.96\right)=\Phi(0.96)-\Phi(-2.96)$$

$$=0.8315-(1-0.9985)=0.83.$$

We find in formula (9.5) that if α is decreased, β will increase. The inverse relation between the Type Ⅰ error and the Type Ⅱ error in Example 9.2 can be generalized. Which means that, for a fixed sample size, if the probability of committing a Type Ⅰ error is decreased, the probability of committing a Type Ⅱ error will increase, and vice versa. In practice, it is usually the case that the Type Ⅰ error is more serious than the Type Ⅱ error. The approach adhered to by most statistical practitioners is to doing their best with respect to type Ⅱ error probability while ensuring that the type Ⅰ error probability is sufficiently small. If we wish to decrease the probability of committing both types of error at the same time, it can be accomplished by increasing the size of the sample.

Sometimes, it is more natural to think about the probability of successfully rejecting H_0 when it is false. This is just $1-P\{\text{Type Ⅱ error}\}=1-P\{\text{Do not reject } H_0 \mid H_0 \text{ is false}\}$, which is known as **the power of the test**[1]. Another way of stating the trade-off is that we want higher power of the test but lower Type Ⅰ error probability.

9.2 Hypotheses on a Single Normal Population

The normal population distribution $N(\mu,\sigma^2)$ is commonly used. The hypothesis testing on the two parameters μ and σ^2 is often encountered in practical application. In this section, we formally consider tests of hypotheses on the parameters of a single normal population.

[1] the power of the test: 检验的势

In the following section, we let $X_1, X_2, \cdots, X_n$ be a random sample from normal population $N(\mu, \sigma^2)$, $x_1, x_2, \cdots, x_n$ be the observed value of this sample. Then, the sample mean and sample variance are as follows, $\overline{X}=\frac{1}{n}\sum_{i=1}^{n}X_i$, $S^2=\frac{1}{n}\sum_{i=1}^{n}(X_i-\overline{X})^2$. The observed value of $\overline{X}$ and S^2 are $\overline{x}=\frac{1}{n}\sum_{i=1}^{n}x_i$, $s^2=\frac{1}{n}\sum_{i=1}^{n}(x_i-\overline{x})^2$.

9.2.1 Hypothesis Concerning a Single Mean

1. σ^2 Known

Consider first the **two-tailed hypothesis**[❶] on sample mean

$$H_0: \mu=\mu_0, \quad H_1: \mu \neq \mu_0$$

Many of the illustrations in previous section 9.1 involve tests on the normal mean μ as above hypothesis. The test statistic is $Z=\frac{\overline{X}-\mu_0}{\sigma/\sqrt{n}}$; The critical region for a fixed significance level α is $|z|=\left|\frac{\overline{x}-\mu_0}{\sigma/\sqrt{n}}\right| \geqslant z_{\alpha/2}$.

Secondly, let's consider test of one-tailed hypothesis on the mean. For giving significance level, we test **right-tailed hypothesis**[❷]

$$H_0: \mu \leqslant \mu_0, \quad H_1: \mu > \mu_0.$$

If the alternative hypothesis $H_1: \mu > \mu_0$ is true, the observed value $\overline{x}$ tends to be large. Therefore, the form of the critical region of right-tailed hypothesis is

$$\overline{x} \geqslant k_0$$

since

$$P\{\overline{X} \geqslant k_0 \mid H_0 \text{ is true}\}=P\left\{\frac{\overline{X}-\mu_0}{\sigma/\sqrt{n}} \geqslant \frac{k_0-\mu_0}{\sigma/\sqrt{n}} \middle| \mu \leqslant \mu_0\right\} \leqslant P\left\{\frac{\overline{X}-\mu}{\sigma/\sqrt{n}} \geqslant \frac{k_0-\mu_0}{\sigma/\sqrt{n}} \middle| \mu \leqslant \mu_0\right\}. \tag{9.6}$$

From the point of view of controlling the probability of committing the Type Ⅰ error, we let $P\left\{\frac{\overline{X}-\mu}{\sigma/\sqrt{n}} \geqslant \frac{k_0-\mu_0}{\sigma/\sqrt{n}} \middle| \mu \leqslant \mu_0\right\}=\alpha$. It can be obtained $\frac{k_0-\mu_0}{\sigma/\sqrt{n}}=z_\alpha$ because of $\frac{\overline{X}-\mu_0}{\sigma/\sqrt{n}} \sim N(0,1)$. That is, the critical region of the right-tailed hypoth-

---

❶ two-tailed hypothesis：双侧检验

❷ right-tailed hypothesis：右侧检验

esis is $z=\dfrac{\overline{x}-\mu_0}{\sigma/\sqrt{n}}\geqslant z_\alpha$.

Similarly, the critical region of the **left-tailed hypothesis**[1] $H_0:\mu\geqslant\mu_0$, $H_1:\mu<\mu_0$ is

$$z=\frac{\overline{x}-\mu_0}{\sigma/\sqrt{n}}\leqslant -z_\alpha.$$

【Example 9. 3】 The mean breaking strength of a certain type of cord has been established from considerable experience at 18. 3 ounces with a standard deviation of 1. 3 ounces. A new machine is purchased to manufacture this type of cord. A sample of 100 pieces obtained from the new machine shows a mean breaking strength of 17. 0 ounces. Would you say that this sample is inferior on the basis of the 1% significance level?

Solution: (1) The test hypothesis is $H_0:\mu\geqslant 18.3$, $H_1:\mu<18.3$.

(2) The appropriate test statistic is $Z=\dfrac{\overline{X}-\mu_0}{\sigma/\sqrt{n}}$.

(3) The critical region is $z\leqslant -z_\alpha$, where $z=\dfrac{\overline{x}-\mu_0}{\sigma/\sqrt{n}}$. Since $\alpha=0.01$, we can find $z_\alpha=2.325$ from appendix Table 2.

(4) Calculations: $\overline{x}=17, \sigma=1.3, n=100$, hence,

$$z=\frac{\overline{x}-\mu_0}{\sigma/\sqrt{n}}=\frac{17-18.3}{1.3/\sqrt{100}}=\frac{-1.3}{0.13}=-10.$$

(5) Decision: Since the observed value of test statistic $z=-10<-z_\alpha=-2.325$, we reject H_0 in favor of H_1, and we conclude that the sample is inferior on the basis of the 1% significance level.

2. σ^2 unknown

In many practical situations where σ^2 is unknown. The statistic $\dfrac{\overline{X}-\mu_0}{\sigma/\sqrt{n}}$ cannot be used to get critical region in this case. We notice that S^2 is an unbiased estimator of σ^2, so, we use S^2 instead of σ^2 to get the test statistic

$$T=\frac{\overline{X}-\mu_0}{S/\sqrt{n}}. \tag{9.7}$$

For the two-sided hypothesis $H_0:\mu=\mu_0$, $H_1:\mu\neq\mu_0$, if the null hypothesis

[1] left-tailed hypothesis: 左侧检验

H_0 is true, the test statistic $T=\dfrac{\overline{X}-\mu_0}{S/\sqrt{n}}\sim t(n-1)$. The structure of the test is identical to that for the case of σ^2 known. Since

$$P\left\{\left|\frac{\overline{X}-\mu_0}{S/\sqrt{n}}\right|\geqslant t_{\alpha/2}(n-1)\right\}=\alpha. \tag{9.8}$$

Then, the critical region is $|t|=\left|\dfrac{\overline{x}-\mu_0}{s/\sqrt{n}}\right|\geqslant t_{\alpha/2}(n-1)$.

For the right-tailed hypothesis $H_0:\mu\leqslant\mu_0$, $H_1:\mu>\mu_0$, the critical region is given by $t\geqslant t_\alpha(n-1)$. For the left-tailed hypothesis $H_0:\mu\geqslant\mu_0$, $H_1:\mu<\mu_0$, the critical region is given by $t\leqslant -t_\alpha(n-1)$.

【Example 9.4】 The Edison Electric Institute has published figures on the number of kilowatt hours used annually by various home appliances. It is claimed that a vacuum cleaner uses an average of 46 kilowatt hours per year. If a random sample of 12 homes included in a planned study indicates that vacuum cleaners use an average of kilowatt hours per year with a sample standard deviation of 11.9 kilowatt hours, does this suggest at the 0.05 significance level that vacuum cleaners use, on average, less than 46 kilowatt hours annually? Assume the population of kilowatt hours to be normal.

Solution: (1) The test hypothesis is $H_0:\mu\geqslant 46$, $H_1:\mu<46$.

(2) The appropriate test statistic is $T=\dfrac{\overline{X}-\mu_0}{S/\sqrt{n}}$.

(3) The critical region is $t\leqslant -t_\alpha(n-1)$, where $t=\dfrac{\overline{x}-\mu_0}{s/\sqrt{n}}$. Since $\alpha=0.05$, we can find $t_\alpha(n-1)=1.796$ from appendix Table 4.

(4) Calculations: $\overline{x}=42, s=11.9$, and $n=12$, hence

$$t=\frac{42-46}{11.9/\sqrt{12}}=-1.16.$$

(5) Decision: Since the observed value of test statistic $t=-1.16>-1.976$, we don't reject H_0, and we conclude that the average number of kilowatt hours used annually by home vacuum cleaners is not significantly less than 46 at the basis of the 5% significance level.

9.2.2 Hypothesis Concerning a Single Variance

In this section, we are first concerned with testing hypothesis

$$H_0:\sigma^2=\sigma_0^2,\ H_1:\sigma^2\neq\sigma_0^2.$$

Since S^2 is an unbiased estimator of σ^2, we can infer whether there is a significant difference between σ^2 and σ_0^2 by the difference between s^2 and σ_0^2. If the null hypothesis H_0 is true, the value of $\frac{s^2}{\sigma_0^2}$ should swing around 1, neither too much greater than 1 nor too much less than 1.

Therefore, if $\frac{s^2}{\sigma_0^2} \leqslant c_1$ or $\frac{s^2}{\sigma_0^2} \geqslant c_2$, we can reject H_0.

Since $\frac{(n-1)S^2}{\sigma_0^2} \sim \chi^2(n-1)$, when H_0 is true, we use the test statistic

$$\chi^2 = \frac{(n-1)S^2}{\sigma_0^2}, \tag{9.9}$$

the critical region should be $\frac{(n-1)s^2}{\sigma_0^2} \leqslant k_1$ or $\frac{(n-1)s^2}{\sigma_0^2} \geqslant k_2$.

The value of k_1 and k_2 can be determined at the basis of significance level α as follows.

$$P\{\text{reject } H_0 | H_0 \text{ is true}\} = P\left\{\left\{\frac{(n-1)S^2}{\sigma_0^2} \leqslant k_1\right\} \cup \left\{\frac{(n-1)S^2}{\sigma_0^2} \geqslant k_2\right\}\right\}$$
$$\leqslant P\left\{\frac{(n-1)S^2}{\sigma_0^2} \leqslant k_1\right\} + P\left\{\frac{(n-1)S^2}{\sigma_0^2} \geqslant k_2\right\} = \alpha. \tag{9.10}$$

Usually, we let $P\left\{\frac{(n-1)S^2}{\sigma_0^2} \leqslant k_1\right\} = P\left\{\frac{(n-1)S^2}{\sigma_0^2} \geqslant k_2\right\} = \frac{\alpha}{2}$, then

$$k_1 = \chi^2_{1-\frac{\alpha}{2}}(n-1), k_2 = \chi^2_{\frac{\alpha}{2}}(n-1). \tag{9.11}$$

Thus, the critical region is $\frac{(n-1)s^2}{\sigma_0^2} \leqslant \chi^2_{1-\frac{\alpha}{2}}(n-1)$ or $\frac{(n-1)s^2}{\sigma_0^2} \geqslant \chi^2_{\frac{\alpha}{2}}(n-1)$.

For the right-tailed hypothesis $H_0: \sigma^2 < \sigma_0^2$, $H_1: \sigma^2 \geqslant \sigma_0^2$, the critical region is given by $\chi^2 = \frac{(n-1)s^2}{\sigma_0^2} \geqslant \chi^2_\alpha(n-1)$. For the left-tailed hypothesis $H_0: \sigma^2 > \sigma_0^2$, $H_1: \sigma^2 \leqslant \sigma_0^2$, the critical region is given by $\chi^2 = \frac{(n-1)s^2}{\sigma_0^2} \leqslant \chi^2_{1-\alpha}(n-1)$. The test hypothesis concerning a single normal population are summarized in Table 9.1.

【Example 9.5】 The packages of grass seed distributed by a certain company is normally distributed with variance $\sigma^2 = 1.2^2$ decagrams. After the improvement of the packaging process, 10 packages were randomly selected with weights: 46.4, 46.1, 45.8, 47.0, 46.1, 45.9, 45.8, 46.9, 45.2 and 46.0. should we believe that the variance of the weights of all such packages of grass seed has changed at

5% significance level?

Solution: (1) The test hypothesis is $H_0: \sigma^2 = 1.2^2$, $H_1: \sigma^2 \neq 1.2^2$.

(2) The appropriate test statistic is $\chi^2 = \dfrac{(n-1)S^2}{\sigma_0^2}$.

(3) The critical region is $\chi^2 \leqslant \chi^2_{1-\frac{\alpha}{2}}(n-1)$ or $\chi^2 \geqslant \chi^2_{\frac{\alpha}{2}}(n-1)$, where

$\chi^2 = \dfrac{(n-1)s^2}{\sigma_0^2}$.

Since $\alpha = 0.05$, we can find $\chi^2_{0.975}(9) = 2.7$ and $\chi^2_{0.025}(9) = 19.023$ from appendix Table 3.

(4) Calculations:
$$s^2 = \frac{\sum_{i=1}^{n} x_i^2 - \frac{1}{n}\left(\sum_{i=1}^{n} x_i\right)^2}{n-1} = \frac{n\sum_{i=1}^{n} x_i^2 - \left(\sum_{i=1}^{n} x_i\right)^2}{n(n-1)}$$
$$= \frac{10 \times (21273.12) - (461.2)^2}{10 \times 9} = 0.286,$$

hence

$$\chi^2 = \frac{9 \times 0.286}{1.2^2} = 1.7875.$$

(5) Decision: Since the observed value of test statistic $\chi^2 = 1.7875 < 2.7$, we reject H_0, and we conclude that the variance of these batteries has changed at the basis of the 5% significance level.

Table 9.1 One-Sample Tests of Hypotheses

	Null hypothesis H_0	Test statistic	Alternative hypothesis H_1	Critical region
σ^2 known	$\mu = \mu_0$ $\mu \leqslant \mu_0$ $\mu \geqslant \mu_0$	$Z = \dfrac{\overline{X} - \mu_0}{\sigma/\sqrt{n}}$	$\mu \neq \mu_0$ $\mu > \mu_0$ $\mu < \mu_0$	$\lvert z \rvert \geqslant z_{\alpha/2}$ $z \geqslant z_\alpha$ $z \leqslant -z_\alpha$
σ^2 unknown	$\mu = \mu_0$ $\mu \leqslant \mu_0$ $\mu \geqslant \mu_0$	$T = \dfrac{\overline{X} - \mu_0}{S/\sqrt{n}}$	$\mu \neq \mu_0$ $\mu > \mu_0$ $\mu < \mu_0$	$\lvert t \rvert \geqslant t_{\alpha/2}(n-1)$ $t \geqslant t_\alpha(n-1)$ $t \leqslant -t_\alpha(n-1)$
μ unknown	$\sigma^2 = \sigma_0^2$ $\sigma^2 \leqslant \sigma_0^2$ $\sigma^2 \geqslant \sigma_0^2$	$\chi^2 = \dfrac{(n-1)S^2}{\sigma_0^2}$	$\sigma^2 \neq \sigma_0^2$ $\sigma^2 > \sigma_0^2$ $\sigma^2 < \sigma_0^2$	$\chi^2 \leqslant \chi^2_{1-\alpha/2}(n-1)$ or $\chi^2 \geqslant \chi^2_{\alpha/2}(n-1)$ $\chi^2 \geqslant \chi^2_\alpha(n-1)$ $\chi^2 \leqslant \chi^2_{1-\alpha}(n-1)$

【Example 9.6】 A manufacturer of car batteries claims that the life of the batteries is normally distributed with a variance equal to 0.9^2 year. If a random sample of 10 has sample variance 1.2^2 years, should we believe that the variance of these batteries is less than 0.9^2 year at the significance level of 5%?

Solution: (1) The test hypothesis is $H_0: \sigma^2 \leqslant 0.9^2$, $H_1: \sigma^2 > 0.9^2$.

(2) The appropriate test statistic is $\chi^2 = \dfrac{(n-1)S^2}{\sigma_0^2}$.

(3) The critical region is $\chi^2 \geqslant \chi_\alpha^2(n-1)$, where $\chi^2 = \dfrac{(n-1)s^2}{\sigma_0^2}$.

Since $\alpha = 0.05$, we can find $\chi_{0.05}^2(9) = 16.919$ from appendix Table 3.

(4) Calculations: $s^2 = 1.2^2$ and $n = 12$, hence

$$\chi^2 = \frac{9 \times 1.2^2}{0.9^2} = 16.0.$$

(5) Decision: Since the observed value of test statistic $\chi^2 = 16.0 < 16.919$, we don't reject H_0, and we conclude that the variance of these batteries is less than 0.9^2 year at the basis of the 5% significance level.

9.3 Two-Sample Tests of Hypotheses

In this section, we consider tests of hypothesis on the parameters of two normal populations. Let $X_1, X_2, \cdots, X_{n_1}$ and $Y_1, Y_2, \cdots, Y_{n_2}$ be two independent random samples drawn separately from normal population $N(\mu_1, \sigma_1{}^2)$ and $N(\mu_2, \sigma_2{}^2)$, $x_1, x_2, \cdots, x_{n_1}$ and $y_1, y_2, \cdots, y_{n_2}$ be the observed values.

9.3.1 Tests on Two Means

Tests concerning two means represent a set of very important analytical tools for the scientist or engineer. The two-sided hypothesis on two means can be written generally as

$$H_0: \mu_1 = \mu_2, \ H_1: \mu_1 \neq \mu_2.$$

1. Known variances

If σ_1^2 and σ_2^2 are known, the test statistic is given by

$$Z = \frac{\overline{X} - \overline{Y}}{\sqrt{\sigma_1^2/n_1 + \sigma_2^2/n_2}}. \tag{9.12}$$

If the null hypothesis H_0 is true, the test statistic $Z = \dfrac{\overline{X} - \overline{Y}}{\sqrt{\sigma_1^2/n_1 + \sigma_2^2/n_2}} \sim N(0,1)$. As derivation of the critical region to one sample, the critical region is

$|z| \geqslant z_{\alpha/2}$ with $z = \dfrac{\bar{x}-\bar{y}}{\sqrt{\sigma_1^2/n_1+\sigma_2^2/n_2}}$ and significance level α. One-tailed critical regions are used in the case of the one-tailed hypothesis. That is, the upper-tailed critical region $z \geqslant z_\alpha$ is to right-tailed alternative hypothesis $H_1: \mu_1 > \mu_2$, and the lower-tailed critical region is to the left-tailed alternative hypothesis $H_1: \mu_1 < \mu_2$.

2. Unknown but equal variances

The more prevalent situations involving tests on two means are those in which variances are unknown. If both distributions are normal and $\sigma_1{}^2 = \sigma_2{}^2 = \sigma^2$, the two-sample T-test may be used. The test statistic is given by

$$T = \frac{\bar{X}-\bar{Y}}{S_W\sqrt{1/n_1+1/n_2}} \tag{9.13}$$

where

$$S_W^2 = \frac{(n_1-1)S_1^2+(n_2-1)S_2^2}{n_1+n_2-2}. \tag{9.14}$$

For the two-tailed hypothesis $H_0: \mu_1 = \mu_2$, $H_1: \mu_1 \neq \mu_2$, we reject H_0 at significance level α when the observed value of T-statistic

$$t = \frac{\bar{x}-\bar{y}}{s_w\sqrt{1/n_1+1/n_2}} \quad \text{where } s_w^2 = \frac{(n_1-1)s_1^2+(n_2-1)s_2^2}{n_1+n_2-2},$$

exceeds $t_{\alpha/2}(n_1+n_2-2)$ or is less than $-t_{\alpha/2}(n_1+n_2-2)$.

One-tailed alternatives suggest one-tailed critical regions as we expect. For example, for right-tailed alternative hypothesis $H_1: \mu_1 > \mu_2$, reject $H_0: \mu_1 \leqslant \mu_2$ when $t \geqslant t_\alpha(n_1+n_2-2)$.

【Example 9.7】 An experiment was performed to compare the abrasive wear of two different laminated materials. 12 pieces of material 1 were tested by exposing each piece to a machine measuring wear. 10 pieces of material 2 were similarly tested. In each case, the depth of wear was observed. The samples of material 1 gave an average wear of 85.5 units with sample standard deviation of 4, while the samples of material 2 gave an average of 81 with a sample standard deviation of 5. Can we conclude at the 0.05 level of significance that the abrasive wear of material 1 is the same as material 2? Assume the populations to be approximately normal with equal variances.

Solution: Let μ_1 and μ_2 respectively represent the population means of the abrasive wear for material 1 and material 2.

(1) The test hypothesis is $H_0: \mu_1 = \mu_2$, $H_1: \mu_1 \neq \mu_2$.

(2) Since the variances are assumed to be equal, the appropriate test statistic is

$$T = \frac{\bar{X}-\bar{Y}}{S_W\sqrt{1/n_1+1/n_2}}.$$

(3) The critical region is $|t| \geqslant t_{\alpha/2}(n_1+n_2-2)$, where $t=\dfrac{\bar{x}-\bar{y}}{s_w\sqrt{1/n_1+1/n_2}}$.

(4) Calculations: $\bar{x}=85.5, s_1=4, n_1=12, \bar{y}=81, s_2=5, n_2=10, \alpha=0.05$. Hence

$$s_w=\sqrt{\frac{11\times16+9\times25}{12+10-2}}=4.478$$

$$t=\frac{85.5-81}{4.478\sqrt{1/12+1/10}}=2.347$$

Since $\alpha=0.05$, we can find $t_{0.025}(20)=2.086$ from appendix Table 4.

(5) Decision: Since the observed value of test statistic $|t|=2.347>2.086$, we reject H_0, and we conclude that the abrasive wear of material 1 is different from material 2 at 5% significance level.

【Example 9.8】 Two methods were used to determine the melting heat from the ice of -0.72℃ to water of 0℃. 13 samples were measured with method A: 79.98, 80.04, 80.02, 80.04, 80.03, 80.03, 80.04, 79.97, 80.05, 80.03, 80.02, 80.00, 80.02, and 8 samples were measured with method B: 80.02, 79.94, 79.98, 79.97, 79.97, 80.03, 79.95, 79.97. Can we conclude at the 0.05 level of significance that the melting heat measured by method A is larger than method B? Assume the populations to be approximately normal with equal variances.

Solution: Let μ_1 and μ_2 respectively represent the population means of the melting heat for method A and method B.

(1) The test hypothesis is $H_0:\mu_1\leqslant\mu_2$, $H_1:\mu_1>\mu_2$.

(2) The test statistic is $T=\dfrac{\overline{X}-\overline{Y}}{S_W\sqrt{1/n_1+1/n_2}}$.

(3) The critical region is $t\geqslant t_\alpha(n_1+n_2-2)$, where $t=\dfrac{\bar{x}-\bar{y}}{s_w\sqrt{1/n_1+1/n_2}}$.

(4) Calculations: $\bar{x}=80.02, s_1^2=0.024^2, n_1=13, \bar{y}=79.98, s_2^2=0.031^2, n_2=8, \alpha=0.05$.

Hence,

$$s_w=\sqrt{\frac{12\times0.024^2+7\times0.031^2}{13+8-2}}=0.0268,$$

$$t=\frac{80.02-79.98}{0.0268\sqrt{1/13+1/8}}=3.333,$$

since $\alpha=0.05$, we can find $t_{0.05}(19)=1.7291$ from appendix Table 4.

(5) Decision: Since $t=3.333>1.7291$, we reject H_0, and we conclude that the melting heat measured by method A is larger than method B at 5% significance level.

3. Unknown and unequal variances

In some cases, the analyst is not able to assume that the variances of two populations are equal. Recall from Section 8.3 that, if the populations are normal, the statistic

$$T'=\frac{(\overline{X}_1-\overline{X}_2)-(\mu_1-\mu_2)}{\sqrt{S_1^2/n_1+S_2^2/n_2}} \tag{9.15}$$

has an approximate t-distribution with approximate degrees of freedom

$$v=\frac{(s_1^2/n_1+s_2^2/n_2)^2}{(s_1^2/n_1)^2/(n_1-1)+(s_2^2/n_2)^2/(n_2-1)}. \tag{9.16}$$

Thus, the critical region to the two-tailed hypothesis $H_0:\mu_1=\mu_2$, $H_1:\mu_1\neq\mu_2$ will be $|t'|\geqslant t_{\alpha/2}(v)$. As the case of equal variances, one-tailed alternatives suggest one-tailed critical regions.

9.3.2 Tests on Two Variances

Methods for comparing two population variances are occasionally needed, though such problems arise much less frequently than those involving means. For the case in which the populations are normal, we usually use the F-test to test the hypothesis.

we consider the hypothesis

$$H_0:\sigma_1^2=\sigma_2^2,\ H_1:\sigma_1^2\neq\sigma_2^2.$$

If the null hypothesis H_0 is true, $E(S_1^2)=\sigma_1^2=\sigma_2^2=E(S_2^2)$. Therefore, we can infer whether there is a significant difference between σ_1^2 and σ_2^2 by the difference between s_1^2 and s_2^2. As the test for single sample variance, the critical region should be $\frac{s_1^2}{s_2^2}\leqslant k_1$ or $\frac{s_1^2}{s_2^2}\geqslant k_2$.

$$\begin{aligned}P\{\text{reject } H_0\mid H_0 \text{ is true}\}&=P\left\{\left\{\frac{S_1^2}{S_2^2}\leqslant k_1\right\}\cup\left\{\frac{S_1^2}{S_2^2}\geqslant k_2\right\}\right\}\\&\leqslant P\left\{\frac{S_1^2}{S_2^2}\leqslant k_1\right\}+P\left\{\frac{S_1^2}{S_2^2}\geqslant k_2\right\}.\end{aligned} \tag{9.17}$$

Recall from Theorem 7.12, we know that, if the sample from two normal populations, the statistic $\frac{S_1^2/S_2^2}{\sigma_1^2/\sigma_2^2}\sim F(n_1-1,n_2-1)$. Thus, the appropriate test statistic will be $F=\frac{S_1^2}{S_2^2}$, and $F=\frac{S_1^2}{S_2^2}\sim F(n_1-1,n_2-1)$ when H_0 is true. let

$$P\left\{\frac{S_1^2}{S_2^2}\leqslant k_1\right\}=P\left\{\frac{S_1^2}{S_2^2}\geqslant k_2\right\}=\frac{\alpha}{2}, \tag{9.18}$$

then $k_1=F_{1-\alpha/2}(n_1-1,n_2-1)$, $k_2=F_{\alpha/2}(n_1-1,n_2-1)$.

Thus, the critical region is $f=\frac{s_1^2}{s_2^2}\leqslant F_{1-\alpha/2}(n_1-1,n_2-1)$ or $f=\frac{s_1^2}{s_2^2}\geqslant F_{\alpha/2}(n_1-1,n_2-1)$. For the right-tailed hypothesis $H_0:\sigma_1^2\leqslant\sigma_2^2$, $H_1:\sigma_1^2>\sigma_2^2$, the critical region is given by $f\geqslant F_\alpha(n_1-1,n_2-1)$. For the left-tailed hypothesis $H_0:\sigma_1^2\geqslant\sigma_2^2$, $H_1:\sigma_1^2<\sigma_2^2$, the critical region is given by $f\leqslant F_{1-\alpha}(n_1-1,n_2-1)$. For convenience, the tests of hypothesis on the parameters of two normal populations are summarized in Table 9.2.

Table 9.2 Two-Sample Tests of Hypotheses

	Null hypothesis H_0	Test statistic	Alternative hypothesis H_1	Critical region
σ_1^2,σ_2^2 known	$\mu_1=\mu_2$ $\mu_1\leqslant\mu_2$ $\mu_1\geqslant\mu_2$	$Z=\dfrac{\overline{X}-\overline{Y}}{\sqrt{\dfrac{\sigma_1^2}{n_1}+\dfrac{\sigma_2^2}{n_2}}}$	$\mu_1\neq\mu_2$ $\mu_1>\mu_2$ $\mu_1<\mu_2$	$\lvert z\rvert\geqslant z_{\alpha/2}$ $z\geqslant z_\alpha$ $z\leqslant -z_\alpha$
σ_1^2,σ_2^2 unknown but equal	$\mu_1=\mu_2$ $\mu_1\leqslant\mu_2$ $\mu_1\geqslant\mu_2$	$T=\dfrac{\overline{X}-\overline{Y}}{S_W\sqrt{\dfrac{1}{n_1}+\dfrac{1}{n_2}}}$ $S_W^2=$ $\dfrac{(n_1-1)S_1^2+(n_2-1)S_2^2}{n_1+n_2-2}$	$\mu_1\neq\mu_2$ $\mu_1>\mu_2$ $\mu_1<\mu_2$	$\lvert t\rvert\geqslant t_{\alpha/2}(n_1+n_2-1)$ $t\geqslant t_\alpha(n_1+n_2-1)$ $t\leqslant -t_\alpha(n_1+n_2-1)$
σ_1^2,σ_2^2 unknown and unequal	$\mu_1=\mu_2$ $\mu_1\leqslant\mu_2$ $\mu_1\geqslant\mu_2$	$Z=\dfrac{\overline{X}-\overline{Y}}{\sqrt{\dfrac{S_1^2}{n_1}+\dfrac{S_2^2}{n_2}}}$	$\mu_1\neq\mu_2$ $\mu_1>\mu_2$ $\mu_1<\mu_2$	$\lvert t\rvert\geqslant t_{\alpha/2}(v)$ $t\geqslant t_\alpha(v)$ $t\leqslant -t_\alpha(v)$ $v=$ $\dfrac{(s_1^2/n_1+s_2^2/n_2)^2}{(s_1^2/n_1)^2/(n_1-1)+(s_2^2/n_2)^2/(n_2-1)}$
μ_1,μ_2 unknown	$\sigma_1^2=\sigma_2^2$ $\sigma_1^2\leqslant\sigma_2^2$ $\sigma_1^2\geqslant\sigma_2^2$	$F=\dfrac{S_1^2}{S_2^2}$	$\sigma_1^2\neq\sigma_2^2$ $\sigma_1^2>\sigma_2^2$ $\sigma_1^2<\sigma_2^2$	$f\leqslant F_{1-\alpha/2}(n_1-1,n_2-1)$ or $f\geqslant F_{\alpha/2}(n_1-1,n_2-1)$ $f\geqslant F_\alpha(n_1-1,n_2-1)$ $f\leqslant F_{1-\alpha}(n_1-1,n_2-1)$

【Example 9.9】 We assume the populations with equal variances in Example 9.8, could you try to test whether our hypothesis is reasonable or not at the 0.05 significance level?

Solution: Let σ_1^2 and σ_2^2 respectively represent the population variances of the melting heat for method A and method B.

(1) The test hypothesis is $H_0:\sigma_1^2=\sigma_2^2$, $H_1:\sigma_1^2\neq\sigma_2^2$.

(2) The test statistic is $F=\dfrac{S_1^2}{S_2^2}$.

(3) The critical region is $f\leqslant F_{1-\alpha/2}(n_1-1,n_2-1)$ or $f\geqslant F_{\alpha/2}(n_1-1,n_2-1)$. $t\geqslant t_\alpha(n_1+n_2-2)$, where $f=\dfrac{s_1^2}{s_2^2}$.

(4) Calculations: $s_1^2=0.024^2$, $n_1=13$, $s_2^2=0.031^2$, $n_2=8$, $\alpha=0.05$, hence

$$f=\frac{s_1^2}{s_2^2}=\frac{0.024^2}{0.031^2}=0.5594,$$

$$F_{0.975}(12,7)=\frac{1}{F_{0.025}(7,12)}=\frac{1}{3.61}=0.277,\ F_{0.025}(12,7)=4.67.$$

(5) Decision: Since $0.277<f=0.5594<4.67$, we can't reject H_0, and we conclude that it is reasonable to assume the populations with equal variances.

Exercises

1. The average service life of an electronic component can not be less than 1000 hours. The producer randomly selects 25 samples from a batch of such components, and the average life of the electronic component is 950 hours. It is known that the life of this element has a normal distribution with a standard deviation of 100. Determine whether these components are qualified at the significance level of 0.05.
2. Experience shows that a fixed dose of a certain drug causes an average increase in pulse rate of 10 beats per minute with a standard deviation of 4.9 patients given the same dose showed the following increases: 13, 15, 14, 10, 8, 12, 16, 9, 20. Could you conclude that the mean increase of this group is significantly different from that of the population at the significance level of 0.05?

   If the population standard deviation of 4 is unknown, solve the problem.
3. A random sample of soil specimens was obtained, and the amount of organic matter (%) in the soil was determined for each specimen. The values of the

sample mean, sample standard deviation, and (estimated) standard error of the mean are 2.481, 1.616, and 0.295, respectively. Does this data suggest that the true average percentage of organic matter in such soil is something other than 3%? Carry out a test of the appropriate hypotheses at significance level of 0.10.

4. According to the regulations, the average content of vitamin C in 100g canned tomato juice should not be less than 21mg. 17 cans have been randomly selected from the products produced by the factory. The content of vitamin C in 100g tomato juice is determined as follows:

   16 25 21 20 23 21 19 15 13 23 17 20 29 18 22 16 22

   Does this indicate that the content of vitamin C in 100g canned tomato juice meet the requirements at the significance level of 0.05? Assuming that the content of vitamin C has a normal distribution.

5. The experience indicates that the time required for high school seniors to complete a standardized test is a normal random variable with a standard deviation $\sigma=6$ minutes. If a random sample of 20 high school seniors has a sample standard deviation $s=4.51$, can we still support the past experience at the significance level of 0.05.

6. The standard deviation of the plant height of a kind of wheat was 12. After purification, 10 plants were randomly selected and their heights were measured as follows:

   90 105 101 95 100 100 101 105 93 97

   Can we conclude that the purified wheat grows more neatly than the original at the significance level of 0.05?

7. Measurements are taken of the lengths (in meters) of adult lizards from two populations separated geographically. The observed values from a sample of size 40 from population 1 can be summarized by $\sum_{i=1}^{40} x_i=14.035$, $\sum_{i=1}^{40} x_i^2=4.937821$, and those from a sample of size 25 from population 2 by $\sum_{i=1}^{25} y_i=8.404$, $\sum_{i=1}^{25} y_i^2=2.833304$. Explain whether the mean adult lengths are the same or not in the two populations at the significance level of 0.05?

8. An eight-week trial of teaching Mathematics to children aged 6 years has been carried out. Those in Group 1 were praised, while the others in Group 2 were not. At the end of the trial all children took an examination. A summary of the examination results is as follows:

Group	Sample size	Sample mean	Sample standard deviation
1	21	51. 48	11. 022
2	23	41. 52	17. 152

Can you conclude that the examination results of group 1 is better than group 2 at the significance level of 0. 05?

9. Strength tests on two types of wool fabric gives the following results (in pounds per square inch):

   Type 1: 138, 127, 134, 125

   Type 2: 124, 137, 175, 140, 130, 134

   Can we conclude that the variance of two samples are homogeneous at the significance level of 0. 05?

10. The following observations are on time (h) for AA 1. 5-volt alkaline battery to reach a 0. 8 voltage.

    Energizer: 8. 65 8. 74 8. 91 8. 72 8. 85 8. 52 8. 62 8. 68 8. 86

    Ultracell: 8. 76 8. 81 8. 81 8. 70 8. 73 8. 76 8. 68 8. 64 8. 79

    Does the data suggest that the variance of the Energizer population distribution differs from that of the Ultracell population distribution at a significance level of 0. 05? Assume that the population distributions are normal.

11. Two machines produce metal parts. Samples with sample sizes of 60 and 40 are taken from the parts produced by the two machines, respectively. The sample variances of component weight are 15 and 9. 66, respectively. Suppose the two populations have normal distribution $N(\mu_1, \sigma_1^2)$ and $N(\mu_2, \sigma_2^2)$ with $\mu_i, \sigma_i^2 (i=1,2)$ unknown. Try to test the hypothesis at significance level 0. 05.

$$H_0: \sigma_1^2 \leqslant \sigma_2^2, \ H_2: \sigma_1^2 > \sigma_2^2.$$

Chapter 10 Application of R in Probability and Statistics

Probability and statistics are a highly applied subject of mathematics. In practical application, we often encounter tedious calculation problems. In this case, statistical analysis software R is very useful. This chapter will introduce the use of R in combination with the examples in probability and statistics.

10.1 R Software Overview

10.1.1 Download and Installation of R Software

The software R is completely free, and the source code is open to everyone. You can download the windows version of R in the website https://www.r-project.org/, the latest version of R is **R version 3.6.1**. The installation of R is very easy, run the program you downloaded and install it according to the prompts of Windows.

The R Console or Command Window (see Figure 10.1) is where commands are typed for immediate execution. The command prompt symbol > is displayed in the Command Window whenever it is waiting for instructions. You type a command and press return. If R thinks that a command you have typed and returned is incomplete, the prompt will change from the usual > to +. You can then add whatever is necessary to complete the command, or press Escape to begin afresh. If you intend to use a series of commands during a session, you should use a script window. it easier to correct any mistakes or make changes to your commands, and you can save your work afterwards. To open a script window, select **File > New script** from the menu bar.

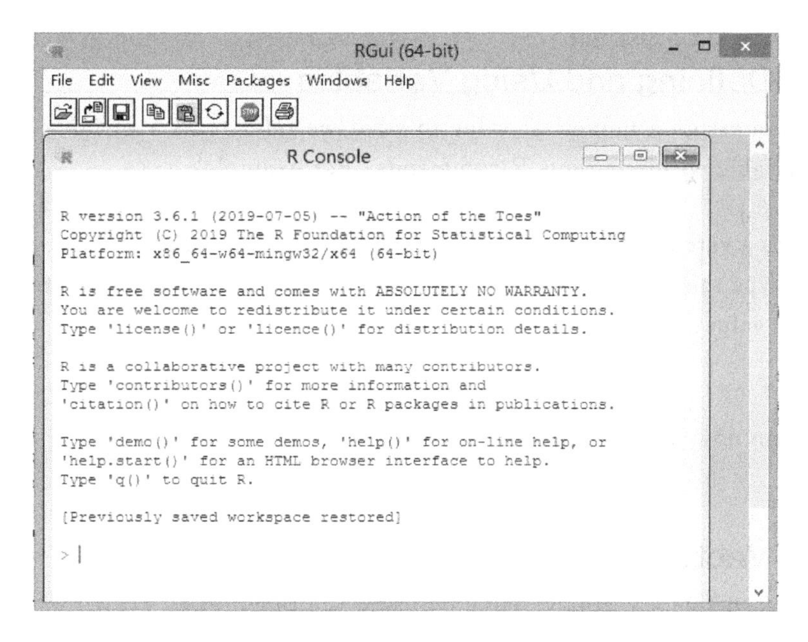

Figure 10. 1 The R Console Window

Help is available online within R. Help on a specific command can be obtained by typing ? **command** or **help(command)**. For example, help on the function mean, which not surprisingly calculates a mean, can be found by typing ? **mean** or **help (mean)**. Information will appear in the Help tab.

10. 1. 2 Using R as a Calculator

To do addition, subtraction, multiplication and division in R use the symbols +, −, * and / respectively. For example:

```
>4*9
[1] 36
```

The symbol ^ is used for raising a number to a power. For example, to calculate 2^{12}:

```
>2^12
[1] 4096
```

You can use many standard mathematical functions such as **sin, cos, tan, log** and **exp**. Note that angles are specified in radians, and the default base for the log command is e. R will interpret the word **pi** as the constant π. Some examples:

```
>cos(2*pi)
[1] 1
>exp(5)
[1] 149.4132
>log(100)
```

```
[1] 4.60517
```

10.1.3 Defining and Using Variables

You can assign a numerical value to what we refer to as a ***variable***, and then use the variable within various R commands. For example:

```
>x<-3
```

Defines a variable called x, which takes the value 3. You won't see any output when you type this command, but if you type the variable name on its own, R will tell you its value:

```
>x
[1] 3
```

You can now use the variable in a command:

```
>2*x
[1] 6
```

10.1.4 Vectors

You can define a vector variable by using the function **c( )**, with a list of the elements separated by commas inside the brackets. For example, to create a vector of the numbers 2, 4, 6, 8, 10, and assign it to a variable x:

```
>x<-c(2,4,6,8,10)
```

If you want to display the vector, type its name.

```
>x
[1] 2 4 6 8 10
```

You can also use vectors within commands.

```
>2*x
[1] 4 8 12 16 20
```

You can select individual elements of vectors, by specifying the element number inside square brackets. For example, to get the fourth element of x, type

```
>x[4]
[1] 8
```

If you define a second vector of the same size, you can do pair-wise operations between the elements of the two vectors. For example:

```
>y<-c(5,4,3,2,1)
>x*y
[1] 10 16 18 18 10
>x^y
[1] 32 256 216 81 10
```

Various commands in R are designed specifically for vectors. For example, to add up all the elements of a vector, use the function **sum( )**:

```
>sum(x)
```

```
[1] 3
```

10.1.5 Plotting Graphs

(1) Scatter Plot

You can do various plots in R by using the function **plot( )**. Here's an example of plotting a cosine curve:

```
>x<-1:10
>plot(x,cos(x))
```

If you try these two commands, you'll see that R has only plotted 10 points (Figure 10.2). If you want to join them up, add the argument type="l" ("l" for line) to the plot command.

```
>plot(x,cos(x),type= "l")
```

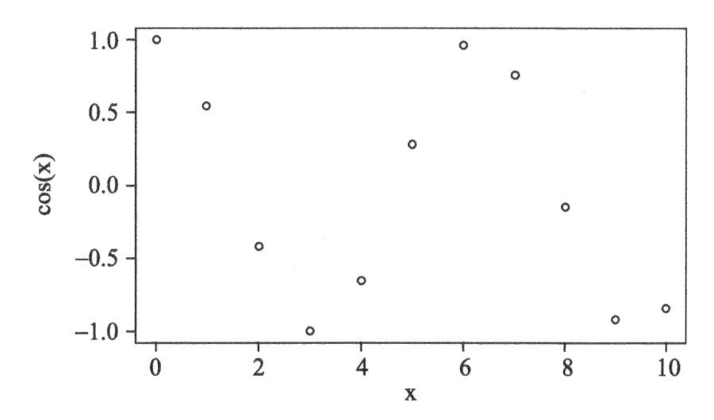

Figure 10.2　The scatter plot of cos(x)

You'll see that the 10 points have been connected by straight lines, which doesn't give a smooth looking curve. To get a smoother looking curve, first define more points on the x-axis:

```
>x<-seq(from= 1,to= 10,length= 100)
```

This creates a sequence of 100 points, evenly spaced, between 1 and 10. Now re-do the plot (Figure 10.3):

```
>plot(x,cos(x),type= "l")
```

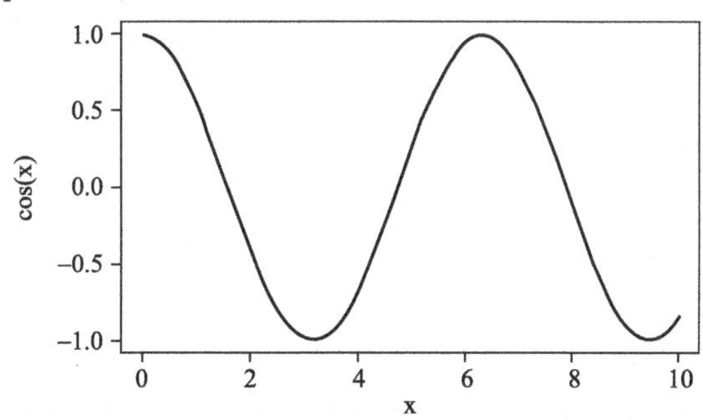

Figure 10.3　The plot of cos(x)

You can add more arguments to have more control over how the plot appears. Run the commands:

```
>x<-seq(from= - pi,to= pi,length= 100)
>plot(x,cos(x),type= "l",lwd= 1,xlab= "x",ylab= "cos(x)",main= "the graph of
cos(x)")
```

The argument lwd refers to the thickness of the line, and increasing the value from 1 will give a thicker line. Try lwd=2. You can specify the axes labels and title by changing the text "x","cos(x)","the graph of cos(x)" to anything you want.

To plot another function on the same axes, use the function lines( ):

```
>lines(x,sin(x))
```

You can change the appearance of the line by specifying a colour, line style, and width:

```
>lines(x,sin(x),col= "red",lty= 2,lwd= 2)
```

(2) Histogram

In R, the histogram of the sample can be drawn by using the function **hist ( )**. Here is an example to draw a histogram of a sample (see Figure 10. 4).

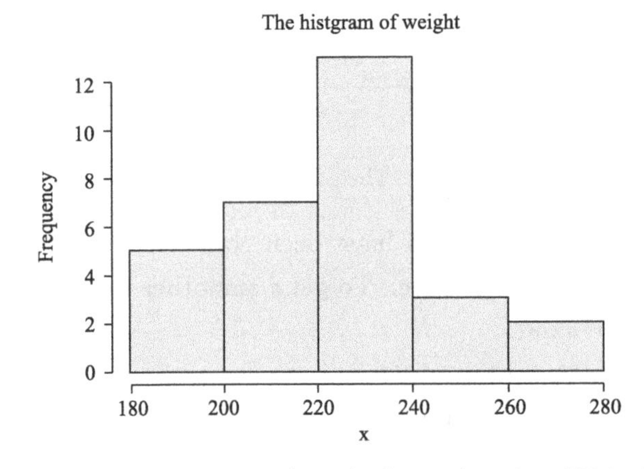

Figure 10. 4 The histogram of weight for 30 American NBA players

Weight data (in pounds) for 30 American NBA players: 225, 232, 232, 245, 235, 245, 270, 225, 240, 240, 217, 195, 225, 185, 200, 220, 200, 210, 271, 240, 220, 230, 215, 252, 225, 220, 206, 185, 227, 236. Run the following command:

```
>x<-c(225,232,232,245,235,245,270,225,240,240,217,195,225,185,
   200,220,200,210,271,240,220,230,215,252,225,220,206,185,227,236)
>hist(x,probability= FALSE,col= "red",main= "The histogram of weight")
```

The argument probability=FALSE refers to the histogram graphic is a representation of frequencies. if probability=TRUE, the histogram graphic is probability densities (so that the histogram has a total area of one).

10.2 R in Solving Probability and Statistical Problems

10.2.1 Probability Calculation

In R, various functions are provided, such as probability density function, cumulative distribution function, quantile, random number and so on.

We introduce these functions by taking the normal distribution as example.

pnorm(*x*, mu, sigma): the value of cumulative distribution function at x for N (mu, sigma);

dnorm(*x*, mu, sigma): the value of probability density function at x for N (mu, sigma);

qnorm(*α*, mu, sigma): Calculate the lower α quantile of the normal distribution N (mu, sigma);

rnorm(*n*, mu, sigma): Generating n random numbers from normal distribution N (mu, sigma).

For other distributions, p, d, q and r are added to the front of the distribution name to represent cumulative distribution function, probability density function, quantile and random number. The names of common distributions in Rare shown in Table 10.1.

Table 10.1 Name of common distribution

Distribution	Name in R	Parameters
Bernoulli distribution Bernoulli (p)	binom	$1, p$
Binomial distribution Bin(n, p)	binom	n, p
Poisson distribution Poisson(λ)	pois	λ
Uniform distribution $U(a, b)$	unif	a, b
Exponential distribution Exp(λ)	exp	λ
Normal distribution $N(\mu, \sigma^2)$	norm	μ, σ
Chi-square distribution $\chi^2(n)$	chisq	n
t-distribution $t(n)$	t	n
F-distribution $F(m, n)$	f	m, n

【Example 10.1】 Let $X \sim N(3, 0.5^2)$,

(1) $P\{X<1\}, P\{X>1.5\}$;

(2) find c to make $P\{X<c\}=0.95$.

R command:

```
>pnorm(1,3,0.5)
[1] 3.167124e-05
>1-pnorm(1.5,3,0.5)
[1] 0.9986501
>qnorm(0.95,3,0.5)
[1] 3.822427
```

10.2.2 Plotting Statistical Graphs

It is a very intuitive and effective to study the common distributions by using the method of plotting. For example, the graph of the probability distribution function of Bin (20, 0.25) (see Figure 10.5) can be drawn by running the command:

```
>x<-0:20
>plot(x,dbinom(x,20,0.25),type= "h",xlab= "x",ylab= "p",main= " The pdf of Bin
(20,0.25) ")
```

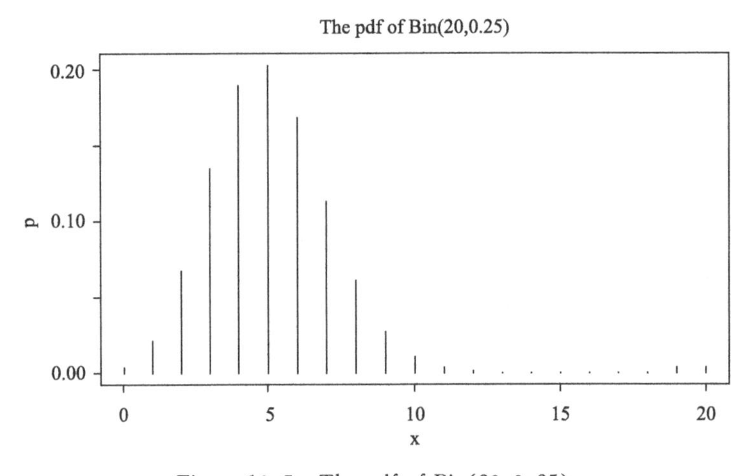

Figure 10.5 The pdf of Bin(20,0.25)

Similarly, we can draw the graph of the probability density function and the cumulative distribution function of N (12, 15^2) (see Figure 10.6, Figure 10.7).

```
>x<-seq(-50,74,1)
>plot(x,dnorm(x,12,15),type= "l",ylab= "f(x}",main= "The pdf of N(12,15^2)")
>plot(x,pnorm(x,12,15),type= "l",ylab= "f(x}",main= "The cdf of N(12,15^2) ")
```

10.2.3 Descriptive Statistics

In R, the function **mean( ), median( ), var( ) and sd( )** can be used separately to calculate the sample mean, sample median, sample variance and standard deviation. For example:

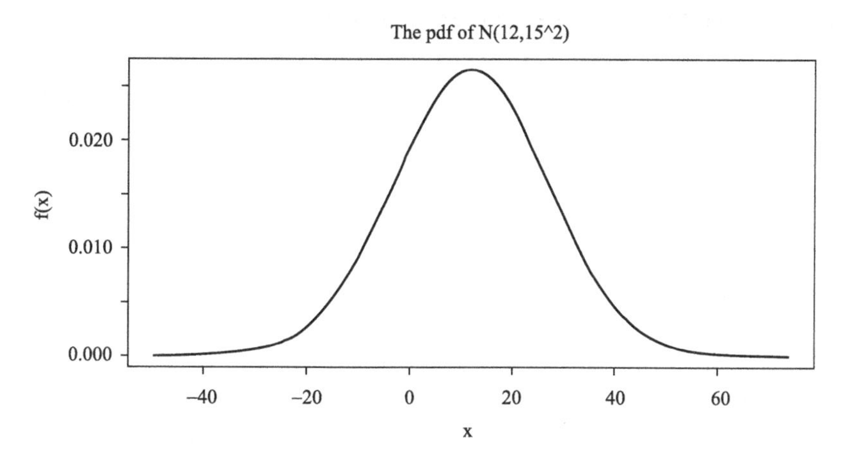

Figure 10.6 The pdf of $N(12,15^2)$

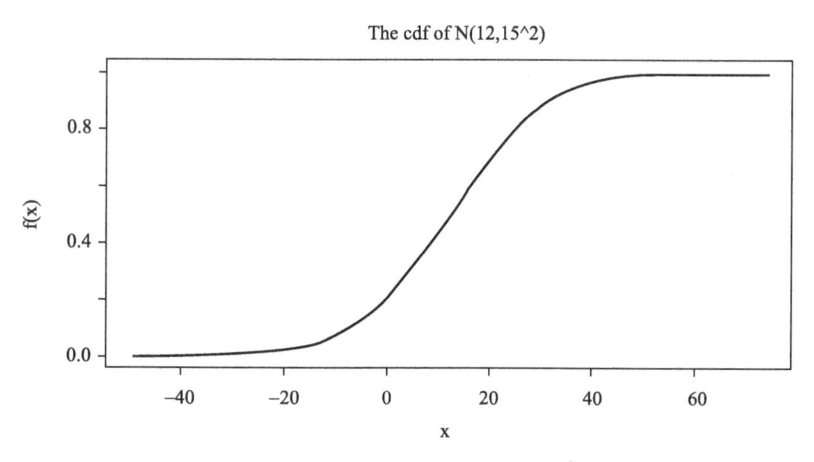

Figure 10.7 The cdf of $N(12,15^2)$

```
>x<-c(45,51,54,61,66,70,74)
>mean(x)
[1] 60.14286
>median(x)
[1] 61
>var(x)
[1] 112.4762
>sd(x)
[1] 10.60548
```

The calculation of k order primitive moment and k order center moment can be obtained by transforming vector x.

```
>mean(x^2)    # 2 order primitive moment
[1] 3713.571
```

```
>mean((x-mean(x))^2)     # 2 order center moment
[1] 96.40816
```

The correlation coefficient between samples can be calculated by the function **cor( )**.

```
>y<-c(100,110,120,130,140,150,160)
>cor(x,y)
[1] 0.996633
```

10.2.4 Estimation in R

1. Method of Moments

There is no special function for moment estimation in R. In fact, as long as the sample moment can be obtained, the moment estimation can be obtained. It is very simple to calculate the k-order origin moment or center moment of the sample.

【Example 10.2】 The distribution density function of a population is

$$f(x,\alpha)=\begin{cases}(\alpha+1)x^{\alpha}, & 0<x<1,\\ 0, & \text{otherwise.}\end{cases}$$

The observations of a sample is 0.1, 0.2, 0.9, 0.8, 0.7, 0.7. Find the moment estimate of α.

Solution: $E(X)=\int_{-\infty}^{+\infty}x\cdot f(x)\mathrm{d}x=\int_{0}^{1}x\cdot(\alpha+1)x^{\alpha}\mathrm{d}x=\frac{\alpha+1}{\alpha+2}x^{\alpha+2}\Big|_{0}^{1}=\frac{\alpha+1}{\alpha+2}$,

Let $\frac{\alpha+1}{\alpha+2}=\overline{X}$, it can be solved $\hat{\alpha}_1=\frac{1}{1-\overline{X}}-2$.

R Command:

```
>x<-c(0.1,0.2,0.9,0.8,0.7,0.7)
>1/(1-mean(x))-2
[1] 0.3076923
```

The moment estimation of α is 0.3076923.

2. Method of Maximum Likelihood

We need to maximize the likelihood function or logarithmic likelihood function in order to find the maximum likelihood estimate. In R, we need to use the function **optimize ( ) or nlm( )**.

The general usage of optimize( ) is

```
optimize(f= ,interval= ,lower= min(interval),upper= max(interval),
         maximum= FALSE,…)
```

The argument f= is the function to be optimized; interval=is a vector containing the end-points of the interval to be searched for the minimum; lower=min (interval) is the lower end point of the interval to be searched; upper=max (interval) is the upper end point of the interval to be searched; maximum = FALSE means to minimize the function, otherwise (maximum=TRUE), to find the maximum.

【Example 10. 3】 Let $(x_1, \cdots, x_n)$ be a set of sample observations drawn from the exponential distribution of parameters λ. Find the estimate of λ by the method of maximum likelihood.

Solution: The logarithmic likelihood function is

$$\ln L(\lambda) = \sum_{i=1}^{n} (\ln\lambda - \lambda x_i) = n\ln\lambda - \lambda \sum_{i=1}^{n} x_i .$$

In order to find the estimate of λ, we should maximize the value of the logarithmic likelihood function above.

You might as well take a random sample of size 20 from an exponential distribution with parameter λ=2.

R Command:

```
>x<-rexp(20,2)
>loglike<-function(lambda) 20*log(lambda)-lambda*sum(x)
>optimize (loglike,c(0,2),maximum= TRUE)
$maximum
[1] 1.999958
$objective
[1] -5.280311
```

The maximum likelihood estimate of λ is 1.999958, which is very close to the actual value 2.

The function **nlm( )** can also be used to find the maximum likelihood estimate. The general usage of nlm( ) is

```
nlm(f,p,…)
```

The argument f is the function to be minimized; p is the starting parameter values for the minimization.

Since the function **nlm( )** is to find the minimum, the minimum of the negative logarithmic likelihood function should be obtained.

The sample taken from the above example can be estimated by function **nlm( )** as follows.

```
>loglike1<-function(lambda) -20*log(lambda)+lambda*sum(x)
>nlm(loglike1,1)
$minimum
[1] 5.2614
$estimate
[1] 2.08951
```

```
$gradient
[1] -3.349516e-07
$code
[1] 1
$iterations
[1] 6
```

The maximum likelihood estimate of λ is 2.08951, which is also very close to the actual value 2.

3. Confidence Intervals for Parameters of Normal Population

(1) Confidence Interval for Single Mean

A $100(1-\alpha)\%$ confidence interval for the mean μ of a normal population with σ^2 known is $\left(\overline{X}-z_{\alpha/2}\cdot\frac{\sigma}{\sqrt{n}},\overline{X}+z_{\alpha/2}\cdot\frac{\sigma}{\sqrt{n}}\right)$; When σ^2 is unknown, the confidence interval will be $\left(\overline{X}-t_{\alpha/2}(n-1)\cdot\frac{S}{\sqrt{n}},\overline{X}+t_{\alpha/2}(n-1)\cdot\frac{S}{\sqrt{n}}\right)$.

【Example 10.4】 (Example 8.15) The contents of seven similar containers of sulfuric acid are 9.8, 10.2, 10.4, 9.8, 10.0, 10.2 and 9.6 liters. Find a 95% confidence interval for the mean contents of all such containers, assuming an approximately normal distribution.

R Command:

```
>x<-c(9.8,10.2,10.4,9.8,10.0,10.2,9.6)
>xu=mean(x)
>n<-length(x)
>xsd<-sd(x)
>q<-qt(1-0.025,n-1)
>LCI<-xu-q*xsd/sqrt(n)
>LCU<-xu+q*xsd/sqrt(n)
>LCI
[1] 9.738414
>LCU
[1] 10.26159
```

A 95% confidence interval for the mean is [9.74, 10.26].

In R, the confidence interval for the mean with σ^2 known can also be obtained by the function **t.test( )**.

```
>x<-c(9.8,10.2,10.4,9.8,10.0,10.2,9.6)
>t.test (x ,conf.level= 0.95)
          One Sample t-test
data:  x
```

```
t= 93.541,df= 6,p-value= 1.006e-10
alternative hypothesis: true mean is not equal to 0
95 percent confidence interval:
  9.738414 10.261586
sample estimates:
mean of x
      10
```

A 95% confidence interval for the mean is [9.738414, 10.261586].

(2) Confidence Intervals for the Difference of the Sample Means

A $100(1-\alpha)\%$ confidence interval for $\mu_1-\mu_2$ with σ_1^2 and σ_2^2 known is

$$\left((\overline{X}_1-\overline{X}_2)-z_{\alpha/2}\sqrt{\frac{\sigma_1^2}{n_1}+\frac{\sigma_2^2}{n_2}},(\overline{X}_1-\overline{X}_2)+z_{\alpha/2}\sqrt{\frac{\sigma_1^2}{n_1}+\frac{\sigma_2^2}{n_2}}\right);$$

If σ_1^2 and σ_2^2 are unknown but equal, the confidence interval is

$$\left((\overline{X}_1-\overline{X}_2)-t_{\alpha/2}(n_1+n_2-2)S_W\sqrt{\frac{1}{n_1}+\frac{1}{n_2}},\right.$$
$$\left.(\overline{X}_1-\overline{X}_2)+t_{\alpha/2}(n_1+n_2-2)S_W\sqrt{\frac{1}{n_1}+\frac{1}{n_2}}\right).$$

【Example 10.5】 Two kinds of rice varieties A and B were sown in 10 experimental fields respectively, and half of them were planted in each experimental field. Assumed that the yield of the two rice varieties have the normal distribution, and the variances are equal. The yields of the 10 experimental fields after harvest are as follows (in kilograms).

A	140	137	136	140	145	148	140	135	144	141
B	135	118	115	140	128	131	130	115	131	125

Find a 95% confidence interval for the difference of the mean between the yield of the two rice varieties.

R Command:

```
>x<-c(140,137,136,140,145,148,140,135,144,141)
>y<-c(135,118,115,140,128,131,130,115,131,125)
>nx<-length(x); ny<-length(y)
>xu<-mean(x); yu<-mean(y)
>varx<-var(x); vary<-var(y)
>sw2<-((nx-1)*varx+(ny-1)*var(y))/(nx+ny-2)
>sw<-sqrt(sw2)
>q<-qt(1-0.025,nx+ny-2)
>LCI<-(xu-yu)-sw*sqrt(1/nx+1/ny)*q
```

```
>LCU<-(xu-yu)+sw*sqrt(1/nx+1/ny)*q
>LCI
[1] 7.536261
>LCU
[1] 20.06374
```

A 95% confidence interval for $\mu_1-\mu_2$ is (7.536261, 20.06374).

The above confidence interval can also be obtained by R Command t.test(x,y, var.equal=TRUE). The argument var.equal=TRUE means the variances are equal.

```
>t.test(x,y,var.equal= TRUE)
          Two Sample t-test
data:  x and y
t= 4.6287,df= 18,p-value= 0.0002087
alternative hypothesis: true difference in means is not equal to 0
95 percent confidence interval:
  7.536261 20.063739
sample estimates:
mean of x mean of y
     140.6     126.8
```

(3) Confidence Intervals for the Variance σ^2

A $100(1-\alpha)\%$ confidence interval for σ^2 of a normal population is

$$\left(\frac{(n-1)S^2}{\chi^2_{\alpha/2}(n-1)},\frac{(n-1)S^2}{\chi^2_{1-\alpha/2}(n-1)}\right).$$

【Example 10.6】 The contents of seven similar containers of sulfuric acid are 9.8, 10.2, 10.4, 9.8, 10.0, 10.2 and 9.6 liters. Find a 95% confidence interval for the variance of the contents of all such containers, assuming an approximately normal distribution.

R Command:

```
>x<-c(9.8,10.2,10.4,9.8,10.0,10.2,9.6)
>xu= mean(x)
>n<-length(x)
>varx<-var(x)
>q1<-qchisq(0.025,n-1)
>q2<-qchisq(1-0.025,n-1)
>LCI<-(n-1)*varx/q2
>LCU<-(n-1)*varx/q1
>LCI
[1] 0.03321943
>LCU
```

```
[1] 0.3879276
```

A 95% confidence interval for the variance is (0.03321943, 0.3879276)

10.2.5 Testing Hypothesis on Mean and Variance of Normal Population

The conclusion of hypothesis test is usually simple. At a given level of significance α, either reject the hypothesis H_0 or accept the hypothesis H_0. However, sometimes it happens that the conclusion of rejecting the original hypothesis H_0 is obtained at a large significance level, but the opposite conclusion is obtained at a small significance level. This problem can be solved by calculating the P-value of hypothesis test.

The P value of the test is the minimum significance level at which observations can be used to reject the original hypothesis H_0. For example, The P value for testing the two-tailed hypothesis $H_0:\mu=\mu_0$, $H_1:\mu\neq\mu_0$ is $P\{|Z|\geqslant z\}$. The conclusion of the test can be easily drawn by comparing the P value (p) with the significance level α. That is, if $p<\alpha$, reject H_0 at significance level α; if $p\geqslant\alpha$, do not reject H_0 at significance level α.

【Example 10.7】 (Example 9.3) The mean breaking strength of a certain type of cord has been established from considerable experience at 18.3 ounces with a standard deviation of 1.3 ounces. A new machine is purchased to manufacture this type of cord. A sample of 100 pieces obtained from the new machine shows a mean breaking strength of 17.0ounces. Would you say that this sample is inferior on the basis of the 1% significance level?

R Command:

```
>mu0<-18.3; sigma<-1.3
>n<-100; xu<-17.0
>z0<-(xu-mu0)/(sigma/sqrt(n))
>q<- -qnorm(0.99,0,1)
>p_value<-pnorm(z0,0,1)
>z0
[1] -10
>q
[1] -2.326348
>p_value
[1] 7.619853e-24
```

Since the observed value of test statistic $z_0=-10<-Z_{0.01}=-2.325$, we reject $H_0:\mu\geqslant 18.3$. On the other hand, because of the p value is 7.619853e-24, it is less than the significance level 0.01, we can draw the same conclusion.

【Example 10.8】 (Example 9.8) Two methods were used to determine the melting heat from the ice of −0.72℃ to water of 0℃. 13 samples were measured with method A: 79.98, 80.04, 80.02, 80.04, 80.03, 80.03, 80.04, 79.97, 80.05, 80.03, 80.02, 80.00, 80.02, and 8 samples were measured with method B: 80.02, 79.94, 79.98, 79.97, 79.97, 80.03, 79.95, 79.97. Can we conclude at the 0.05 level of significance that the melting heat measured by method A is larger than method B? Assume the populations to be approximately normal with equal variances.

R Command:

```
>xa <-c (79.98, 80.04, 80.02, 80.04, 80.03, 80.03, 80.04, 79.97, 80.05, 80.03, 80.02,
80.00,80.02)
>xb<-c(80.02,79.94,79.98,79.97,79.97,80.03,79.95,79.97)
>na<-length(xa); nb<-length(xb)
>xau<-mean(xa); xbu<-mean(xb)
>varxa<-var(xa); varxb<-var(xb)
>sdxa<-sd(xa); sdxb<-sd(xb)
>sw2<-((na-1)*varxa+(nb-1)*var(xb))/(na+nb-2)
>sw<-sqrt(sw2)
>q<-qt(1-0.05,na+nb-2)
>t0<-(xau-xbu)/(sw*sqrt(1/na+1/nb))
>P_value<-1-pt(t0,na+nb-2)
>t0; q; P_value
[1] 3.472245
[1] 1.729133
[1] 0.001275502
```

Since $t_0 = 3.4722 > 1.7291$, we reject $H_0: \mu_1 \leqslant \mu_2$, and we conclude that the melting heat measured by method A is larger than method B at 5% significance level. The same conclusion can be drawn by the P _ value.

In R, the function t. test( ) can be used directly to perform one and two sample T-test.

```
>t.test(xa,xb,alternative= "greater",var.equal= TRUE)
            Two Sample t-test
data:  xa and xb
t= 3.4722,df= 19,p-value= 0.001276
alternative hypothesis: true difference in means is greater than 0
95 percent confidence interval:
  0.0210942     Inf
sample estimates:
mean of x mean of y
  80.02077  79.97875
```

Since the p-value=0.001276<0.05, we reject H_0.

【Example 10.9】 (Example 9.5) The packages of grass seed distributed by a certain company is normally distributed with variance $\sigma^2 = 1.2^2$ decagrams. After the improvement of the packaging process, 10 packages were randomly selected with weights: 46.4, 46.1, 45.8, 47.0, 46.1, 45.9, 45.8, 46.9, 45.2 and 46.0. should we believe that the variance of the weights of all such packages of grass seed has changed at 5% significance level?

R Command:

```
>x<-c(46.4,46.1,45.8,47.0,46.1,45.9,45.8,46.9,45.2,46.0)
>n<-length(x);sigma<-1.2
>varx<-var(x)
>chisq0<-(n-1)*varx/sigma^2
>q1<-qchisq(0.025,n-1);q2<-qchisq(1-0.025,n-1)
>chisq0;q1;q2
[1] 1.788889
[1] 2.700389
[1] 19.02277
```

Since the observed value of test statistic $\chi^2 = 1.7889 < 2.7$, we reject $H_0: \sigma^2 = 1.2^2$, and we conclude that the variance of these batteries has changed at the basis of the 5% significance level.

【Example 10.10】 (Example 9.9) We assume the populations with equal variances in Example 9.9, could you try to test whether our hypothesis is reasonable or not at the 0.05 significance level?

Solution: Let σ_A^2 and σ_B^2 respectively represent the population variances of the melting heat for method A and method B.

R Command:

```
>na<-12; nb<-8
>sa2<-0.024^2; sb2<-0.031^2
>f<-sa2/sb2
>q1<-qf(0.025,12,7); q2<-qf(0.975,12,7)
>f; q1; q2
[1] 0.5993757
[1] 0.277276
[1] 4.66583
```

Since $0.2773 < f = 0.5594 < 4.6658$, we can't reject H_0, and we conclude that it is reasonable to assume the populations with equal variances.

Appendix Statistical Tables

Table 1 Poisson Distribution

$$P\{X \leqslant x\} = \sum_{k=0}^{x} \frac{\lambda^k e^{-\lambda}}{k!}$$

k \ λ	0.1	0.2	0.3	0.4	0.5	0.6	0.7	0.8	0.9
0	0.9048	0.8187	0.7408	0.6703	0.6065	0.5488	0.4966	0.4493	0.4066
1	0.9953	0.9825	0.9631	0.9384	0.9098	0.8781	0.8442	0.8088	0.7725
2	0.9998	0.9989	0.9964	0.9921	0.9856	0.9769	0.9659	0.9526	0.9371
3	1.0000	0.9999	0.9997	0.9992	0.9982	0.9966	0.9942	0.9909	0.9865
4		1.0000	1.0000	0.9999	0.9998	0.9996	0.9992	0.9986	0.9977
5				1.0000	1.0000	1.0000	0.9999	0.9998	0.9997
6							1.0000	1.0000	1.0000

k \ λ	1.0	1.5	2.0	2.5	3.0	3.5	4.0	4.5	5.0
0	0.3679	0.2231	0.1353	0.0821	0.0498	0.0302	0.0183	0.0111	0.0067
1	0.7358	0.5578	0.4060	0.2873	0.1991	0.1359	0.0916	0.0611	0.0404
2	0.9197	0.8088	0.6767	0.5438	0.4232	0.3208	0.2381	0.1736	0.1247
3	0.9810	0.9344	0.8571	0.7576	0.6472	0.5366	0.4335	0.3423	0.2650
4	0.9963	0.9814	0.9473	0.8912	0.8153	0.7254	0.6288	0.5321	0.4405
5	0.9994	0.9955	0.9834	0.9580	0.9161	0.8576	0.7851	0.7029	0.6160
6	0.9999	0.9991	0.9955	0.9858	0.9665	0.9347	0.8893	0.8311	0.7622
7	1.0000	0.9998	0.9989	0.9958	0.9881	0.9733	0.9489	0.9134	0.8666
8		1.0000	0.9998	0.9989	0.9962	0.9901	0.9786	0.9597	0.9319
9			1.0000	0.9997	0.9989	0.9967	0.9919	0.9829	0.9682
10				0.9999	0.9997	0.9990	0.9972	0.9933	0.9863

continued

k \ λ	1.0	1.5	2.0	2.5	3.0	3.5	4.0	4.5	5.0
11				1.0000	0.9999	0.9997	0.9991	0.9976	0.9945
12					1.0000	0.9999	0.9997	0.9992	0.9980
13						1.0000	0.9999	0.9997	0.9993
14							1.0000	0.9999	0.9998
15								1.0000	0.9999
16									1.0000

k \ λ	5.5	6.0	6.5	7.0	7.5	8.0	8.5	9.0	9.5
0	0.0041	0.0025	0.0015	0.0009	0.0006	0.0003	0.0002	0.0001	0.0001
1	0.0266	0.0174	0.0113	0.0073	0.0047	0.0030	0.0019	0.0012	0.0008
2	0.0884	0.0620	0.0430	0.0296	0.0203	0.0138	0.0093	0.0062	0.0042
3	0.2017	0.1512	0.1118	0.0818	0.0591	0.0424	0.0301	0.0212	0.0149
4	0.3575	0.2851	0.2237	0.1730	0.1321	0.0996	0.0744	0.0550	0.0403
5	0.5289	0.4457	0.3690	0.3007	0.2414	0.1912	0.1496	0.1157	0.0885
6	0.6860	0.6063	0.5265	0.4497	0.3782	0.3134	0.2562	0.2068	0.1649
7	0.8095	0.7440	0.6728	0.5987	0.5246	0.4530	0.3856	0.3239	0.2687
8	0.8944	0.8472	0.7916	0.7291	0.6620	0.5925	0.5231	0.4557	0.3918
9	0.9462	0.9161	0.8774	0.8305	0.7764	0.7166	0.6530	0.5874	0.5218
10	0.9747	0.9574	0.9332	0.9015	0.8622	0.8159	0.7634	0.7060	0.6453
11	0.9890	0.9799	0.9661	0.9467	0.9208	0.8881	0.8487	0.8030	0.7520
12	0.9955	0.9912	0.9840	0.9730	0.9573	0.9362	0.9091	0.8758	0.8364
13	0.9983	0.9964	0.9929	0.9872	0.9784	0.9658	0.9486	0.9261	0.8981
14	0.9994	0.9986	0.9970	0.9943	0.9897	0.9827	0.9726	0.9585	0.9400
15	0.9998	0.9995	0.9988	0.9976	0.9954	0.9918	0.9862	0.9780	0.9665
16	0.9999	0.9998	0.9996	0.9990	0.9980	0.9963	0.9934	0.9889	0.9823
17	1.0000	0.9999	0.9998	0.9996	0.9992	0.9984	0.9970	0.9947	0.9911
18		1.0000	0.9999	0.9999	0.9997	0.9993	0.9987	0.9976	0.9957
19			1.0000	1.0000	0.9999	0.9997	0.9995	0.9989	0.9980
20					1.0000	0.9999	0.9998	0.9996	0.9991
21						1.0000	0.9999	0.9998	0.9996
22							1.0000	0.9999	0.9999
23								1.0000	0.9999

Table 2 Standard Normal Distribution Function

$$\Phi(x)=\int_{-\infty}^{x}\frac{1}{\sqrt{2\pi}}e^{-\frac{t^2}{2}}dt$$

x	0.00	0.01	0.02	0.03	0.04	0.05	0.06	0.07	0.08	0.09
0.0	0.5000	0.5040	0.5080	0.5120	0.5160	0.5199	0.5239	0.5279	0.5319	0.5359
0.1	0.5398	0.5438	0.5478	0.5517	0.5557	0.5596	0.5636	0.5675	0.5714	0.5753
0.2	0.5793	0.5832	0.5871	0.5910	0.5948	0.5987	0.6026	0.6064	0.6103	0.6141
0.3	0.6179	0.6217	0.6255	0.6293	0.6331	0.6368	0.6406	0.6443	0.6480	0.6517
0.4	0.6554	0.6591	0.6628	0.6664	0.6700	0.6736	0.6772	0.6808	0.6844	0.6879
0.5	0.6915	0.6950	0.6985	0.7019	0.7054	0.7088	0.7123	0.7157	0.7190	0.7224
0.6	0.7257	0.7291	0.7324	0.7357	0.7389	0.7422	0.7454	0.7486	0.7517	0.7549
0.7	0.7580	0.7611	0.7642	0.7673	0.7703	0.7734	0.7764	0.7794	0.7823	0.7582
0.8	0.7881	0.7910	0.7939	0.7967	0.7995	0.8023	0.8051	0.8078	0.8106	0.8133
0.9	0.8159	0.8186	0.8212	0.8238	0.8264	0.8289	0.8315	0.8340	0.8365	0.8389
1.0	0.8413	0.8438	0.8461	0.8485	0.8508	0.8531	0.8554	0.8577	0.8599	0.8621
1.1	0.8643	0.8665	0.8686	0.8708	0.8729	0.8749	0.8770	0.8790	0.8810	0.8830
1.2	0.8849	0.8869	0.8888	0.8907	0.8925	0.8944	0.8962	0.8980	0.8997	0.9015
1.3	0.9032	0.9049	0.9066	0.9082	0.9099	0.9115	0.9131	0.9147	0.9162	0.9177
1.4	0.9192	0.9207	0.9222	0.9236	0.9251	0.9265	0.9278	0.9292	0.9306	0.9319
1.5	0.9332	0.9345	0.9357	0.9370	0.9382	0.9394	0.9406	0.9418	0.9430	0.9441
1.6	0.9452	0.9463	0.9474	0.9484	0.9495	0.9505	0.9515	0.9525	0.9535	0.9545
1.7	0.9554	0.9564	0.9573	0.9582	0.9591	0.9599	0.9608	0.9616	0.9625	0.9633
1.8	0.9641	0.9648	0.9656	0.9664	0.9671	0.9678	0.9686	0.9693	0.9700	0.9706
1.9	0.9713	0.9719	0.9726	0.9732	0.9738	0.9744	0.9750	0.9756	0.9762	0.9767
2.0	0.9772	0.9778	0.9783	0.9788	0.9793	0.9798	0.9803	0.9808	0.9812	0.9817
2.1	0.9821	0.9826	0.9830	0.9834	0.9838	0.9842	0.9846	0.9850	0.9854	0.9857
2.2	0.9861	0.9864	0.9868	0.9871	0.9874	0.9878	0.9881	0.9884	0.9887	0.9890
2.3	0.9893	0.9896	0.9898	0.9901	0.9904	0.9906	0.9909	0.9911	0.9913	0.9916
2.4	0.9918	0.9920	0.9922	0.9925	0.9927	0.9929	0.9931	0.9932	0.9934	0.9936
2.5	0.9938	0.9940	0.9941	0.9943	0.9945	0.9946	0.9948	0.9949	0.9951	0.9952
2.6	0.9953	0.9955	0.9956	0.9957	0.9959	0.9960	0.9961	0.9962	0.9963	0.9964
2.7	0.9965	0.9966	0.9967	0.9968	0.9969	0.9970	0.9971	0.9972	0.9973	0.9974
2.8	0.9974	0.9975	0.9976	0.9977	0.9977	0.9978	0.9979	0.9979	0.9980	0.9981
2.9	0.9981	0.9982	0.9982	0.9983	0.9984	0.9984	0.9985	0.9985	0.9986	0.9986
3.0	0.9987	0.9990	0.9993	0.9995	0.9997	0.9998	0.9998	0.9999	0.9999	1.0000

Note: The values of the last row in the table are $\Phi(3.0), \Phi(3.1), \cdots, \Phi(3.9)$.

Table 3 Values of χ_{α}^{2}

$P\{\chi^2(n) > \chi_{\alpha}^{2}(n)\} = \alpha$

n	$\alpha=0.995$	0.99	0.975	0.95	0.90	0.75
1	—	—	0.001	0.004	0.016	0.102
2	0.010	0.020	0.051	0.103	0.211	0.575
3	0.072	0.115	0.216	0.352	0.584	1.213
4	0.207	0.297	0.484	0.711	1.064	1.923
5	0.412	0.554	0.831	1.145	1.610	2.675
6	0.676	0.872	1.237	1.635	2.204	3.455
7	0.989	1.239	1.690	2.167	2.833	4.255
8	1.344	1.646	2.180	2.733	3.490	5.071
9	1.735	2.088	2.700	3.325	4.168	5.899
10	2.156	2.558	3.247	3.940	4.865	6.737
11	2.603	3.053	3.816	4.575	5.578	7.584
12	3.074	3.571	4.404	5.226	6.304	8.438
13	3.565	4.107	5.009	5.892	7.042	9.299
14	4.075	4.660	5.629	6.571	7.790	10.165
15	4.601	5.229	6.262	7.261	8.547	11.037
16	5.142	5.812	6.908	7.962	9.312	11.912
17	5.697	6.408	7.564	8.672	10.085	12.792
18	6.265	7.015	8.231	9.390	10.865	13.675
19	6.844	7.633	8.907	10.117	11.651	14.562
20	7.434	8.260	9.591	10.851	12.443	15.452
21	8.034	8.897	10.283	11.591	13.240	16.344
22	8.643	9.542	10.982	12.338	14.042	17.240
23	9.260	10.196	11.689	13.091	14.848	18.137
24	9.886	10.856	12.401	13.848	15.659	19.037
25	10.520	11.524	13.120	14.611	16.473	19.939
26	11.160	12.198	13.844	15.379	17.292	20.843
27	11.808	12.879	14.573	16.151	18.114	21.749
28	12.461	13.565	15.308	16.928	18.939	22.657
29	13.121	14.257	16.047	17.708	19.768	23.567
30	13.787	14.954	16.791	18.493	20.599	24.478
31	14.458	15.655	17.539	19.281	21.434	25.390
32	15.134	16.362	18.291	20.072	22.271	26.304
33	15.815	17.074	19.047	20.867	23.110	27.219
34	16.501	17.789	19.806	21.664	23.952	28.186
35	17.192	18.509	20.569	22.465	24.797	29.054
36	17.887	19.233	21.336	23.269	25.643	29.973
37	18.586	19.960	22.106	24.075	26.492	30.893
38	19.289	20.691	22.878	24.884	27.343	31.815
39	19.996	21.426	23.654	25.695	28.196	32.737
40	20.707	22.164	24.433	26.509	29.051	33.660
41	21.421	22.906	25.215	27.326	29.907	34.585
42	22.138	23.650	25.999	28.144	30.765	35.510
43	22.859	24.398	26.785	28.965	31.625	36.436
44	23.584	25.148	27.575	29.787	32.487	37.363
45	24.311	25.901	28.366	30.612	33.350	38.291

continued

n	α=0.25	0.10	0.05	0.025	0.01	0.005
1	1.323	2.706	3.841	5.024	6.635	7.879
2	2.773	4.605	5.991	7.378	9.210	10.597
3	4.108	6.251	7.815	9.348	11.345	12.838
4	5.385	7.779	9.488	11.143	13.277	14.860
5	6.626	9.236	11.071	12.833	15.086	16.750
6	7.841	10.645	12.592	14.449	16.812	18.548
7	9.037	12.017	14.067	16.013	18.475	20.278
8	10.219	13.362	15.507	17.535	20.090	21.955
9	11.389	14.684	16.919	19.023	21.666	23.589
10	12.549	15.987	18.307	20.483	23.209	25.188
11	13.701	17.275	19.675	21.920	24.725	26.757
12	14.845	18.549	21.026	23.337	26.217	28.299
13	15.984	19.812	22.362	24.736	27.688	29.819
14	17.117	21.064	23.685	26.119	29.141	31.319
15	18.245	22.307	24.996	27.488	30.578	32.801
16	19.369	23.542	26.296	28.845	32.000	34.267
17	20.489	24.769	27.587	30.191	33.409	35.718
18	21.605	25.989	28.869	31.526	34.805	37.156
19	22.718	27.204	30.144	32.852	36.191	38.582
20	23.828	28.412	31.410	34.170	37.566	39.997
21	24.935	29.615	32.671	35.479	38.932	41.401
22	26.039	30.813	33.924	36.781	40.289	42.796
23	27.141	32.007	35.172	38.076	41.638	44.181
24	28.241	33.196	36.415	39.364	42.980	45.559
25	29.339	34.382	37.652	40.646	44.314	46.928
26	30.435	35.563	38.885	41.923	45.642	48.290
27	31.528	36.741	40.113	43.194	46.963	49.645
28	32.620	37.916	41.337	44.461	48.278	50.993
29	33.711	39.087	42.557	45.722	49.588	52.336
30	34.800	40.256	43.773	46.979	50.892	53.672
31	35.887	41.422	44.985	48.232	52.191	55.003
32	36.973	42.585	46.194	49.480	53.486	56.328
33	38.058	43.745	47.400	50.725	54.776	57.648
34	39.141	44.903	48.602	51.966	56.061	58.964
35	40.223	46.059	49.802	53.203	57.342	60.275
36	41.304	47.212	50.998	54.437	58.619	61.581
37	42.383	48.363	52.192	55.668	59.892	62.883
38	43.462	49.513	53.384	56.896	61.162	64.181
39	44.539	50.660	54.572	58.120	62.428	65.476
40	45.616	51.805	55.758	59.342	63.691	66.766
41	46.692	52.949	56.942	60.561	64.950	68.053
42	47.766	54.090	58.124	61.777	66.206	69.336
43	48.840	55.230	59.304	62.990	67.459	70.616
44	49.913	56.369	60.481	64.201	68.710	71.893
45	50.985	57.505	61.656	35.410	69.957	73.166

Table 4 Values of t_{α}

$P\{t(n) > t_{\alpha}(n)\} = \alpha$

n	α=0.25	0.10	0.05	0.025	0.01	0.005
1	1.0000	3.0777	6.3138	12.7062	31.8207	63.6574
2	0.8165	1.8856	2.9200	4.3027	6.9646	9.9248
3	0.7649	1.6377	2.3534	3.1824	4.5407	5.8409
4	0.7407	0.5332	2.1318	2.7764	3.7469	4.6041
5	0.7267	1.4759	2.0150	2.5706	3.3649	4.0322
6	0.7176	1.4398	1.9432	2.4469	3.1427	3.7074
7	0.7111	1.4149	1.8946	2.3646	2.9980	3.4995
8	0.7064	1.3968	1.8595	2.3060	2.8965	3.3554
9	0.7027	1.3830	1.8331	2.2622	2.8214	3.2498
10	0.6998	1.3722	1.8125	2.2281	2.7638	3.1693
11	0.6974	1.3634	1.7959	2.2010	2.7181	3.1058
12	0.6955	1.3562	1.7823	2.1788	2.6810	3.0545
13	0.6938	1.3502	1.7709	2.1604	2.6503	3.0123
14	0.6924	1.3450	1.7613	2.1448	2.6245	2.9768
15	0.6912	1.3406	1.7531	2.1315	2.6025	2.9467
16	0.6901	1.3368	1.7459	2.1199	2.5835	2.9208
17	0.6892	1.3334	1.7396	2.1098	2.5669	2.8982
18	0.6884	1.3304	1.7341	2.1009	2.5524	2.8784
19	0.6876	1.3277	1.7291	2.0930	2.5395	2.8609
20	0.6870	1.3253	1.7247	2.0860	2.5280	2.8453
21	0.6864	1.3232	1.7207	2.0796	2.5177	2.8314
22	0.6858	1.3212	1.7171	2.0739	2.5083	2.8188
23	0.6853	1.3195	1.7139	2.0687	2.4999	2.8073
24	0.6848	1.3178	1.7109	2.0639	2.4922	2.7969
25	0.6844	1.3163	1.7081	2.0595	2.4851	2.7874
26	0.6840	1.3150	1.7056	2.0555	2.4786	2.7787
27	0.6837	1.3137	1.7033	2.0518	2.4727	2.7707
28	0.6834	1.3125	1.7011	2.0484	2.4641	2.7633
29	0.6830	1.3114	1.6991	2.0452	2.4620	2.7564
30	0.6828	1.3104	1.6973	2.0423	2.4573	2.7500
31	0.6825	1.3095	1.6955	2.0395	2.4528	2.7440
32	0.6822	1.3086	1.6939	2.0369	2.4487	2.7385
33	0.6820	1.3077	1.6924	2.0345	2.4448	2.7333
34	0.6818	1.3070	1.6909	2.0322	2.4411	2.7284
35	0.6816	1.3062	1.6896	2.0301	2.4377	2.7238
36	0.6814	1.3055	1.6883	2.0281	2.4345	2.7195
37	0.6812	1.3049	1.6871	2.0262	2.4314	2.7154
38	0.6810	1.3042	1.6860	2.0244	2.4286	2.7116
39	0.6808	1.3036	1.6849	2.0227	2.4258	2.7079
40	0.6807	1.3031	1.6839	2.0211	2.4233	2.7045
41	0.6805	1.3025	1.6829	2.0195	2.4208	2.7012
42	0.6804	1.3020	1.6820	2.0181	2.4185	2.6981
43	0.6802	1.3016	1.6811	2.0167	2.4163	2.6951
44	0.6801	1.3011	1.6802	2.0154	2.4141	2.6923
45	0.6800	1.3006	1.6794	2.0141	2.4121	2.6896

Table 5 Values of F_α

$P\{F(n_1,n_2)>F_\alpha(n_1,n_2)\}=\alpha$

$\alpha=0.10$

n_2 \ n_1	1	2	3	4	5	6	7	8	9	10	12	15	20	24	30	40	60	120	∞
1	39.86	49.50	53.59	55.83	57.24	58.20	58.91	59.44	59.86	60.19	60.71	61.22	61.74	62.00	62.26	62.53	62.79	63.06	63.33
2	8.53	9.00	9.16	9.24	9.29	9.33	9.35	9.37	9.38	9.39	9.41	9.42	9.44	9.45	9.46	9.47	9.47	9.48	9.49
3	5.54	5.46	5.39	5.34	5.31	5.28	5.27	5.25	5.24	5.23	5.22	5.20	5.18	5.18	5.17	5.16	5.15	5.14	5.13
4	4.54	4.32	4.19	4.11	4.05	4.01	3.98	3.95	3.94	3.92	3.90	3.87	3.84	3.83	3.82	3.80	3.79	3.78	3.76
5	4.06	3.78	3.62	3.52	3.45	3.40	3.37	3.34	3.32	3.30	3.27	3.24	3.21	3.19	3.17	3.16	3.14	3.12	3.10
6	3.78	3.46	3.29	3.18	3.11	3.05	3.01	2.98	2.96	2.94	2.90	2.87	2.84	2.82	2.80	2.78	2.76	2.74	2.72
7	3.59	3.26	3.07	2.96	2.88	2.83	2.78	2.75	2.72	2.70	2.67	2.63	2.59	2.58	2.56	2.54	2.51	2.49	2.47
8	3.46	3.11	2.92	2.81	2.73	2.67	2.62	2.59	2.56	2.54	2.50	2.46	2.42	2.40	2.38	2.36	2.34	2.32	2.29
9	3.36	3.01	2.81	2.69	2.61	2.55	2.51	2.47	2.44	2.42	2.38	2.34	2.30	2.28	2.25	2.23	2.21	2.18	2.16
10	3.29	2.92	2.73	2.61	2.52	2.46	2.41	2.38	2.35	2.32	2.28	2.24	2.20	2.18	2.16	2.13	2.11	2.08	2.06
11	3.23	2.86	2.66	2.54	2.45	2.39	2.34	2.30	2.27	2.25	2.21	2.17	2.12	2.10	2.08	2.05	2.03	2.00	1.97
12	3.18	2.81	2.61	2.48	2.39	2.33	2.28	2.24	2.21	2.19	2.15	2.10	2.06	2.04	2.01	1.99	1.96	1.93	1.90
13	3.14	2.76	2.56	2.43	2.35	2.28	2.23	2.20	2.16	2.14	2.10	2.05	2.01	1.98	1.96	1.93	1.90	1.88	1.85
14	3.10	2.73	2.52	2.39	2.31	2.24	2.19	2.15	2.12	2.10	2.05	2.01	1.96	1.94	1.91	1.89	1.86	1.83	1.80
15	3.07	2.70	2.49	2.36	2.27	2.21	2.16	2.12	2.09	2.06	2.02	1.97	1.92	1.90	1.87	1.85	1.82	1.79	1.76
16	3.05	2.67	2.46	2.33	2.24	2.18	2.13	2.09	2.06	2.03	1.99	1.94	1.89	1.87	1.84	1.81	1.78	1.75	1.72
17	3.03	2.64	2.44	2.31	2.22	2.15	2.10	2.06	2.03	2.00	1.96	1.91	1.86	1.84	1.81	1.78	1.75	1.72	1.69
18	3.01	2.62	2.42	2.29	2.20	2.13	2.08	2.04	2.00	1.98	1.93	1.89	1.84	1.81	1.78	1.75	1.72	1.69	1.66
19	2.99	2.61	2.40	2.27	2.18	2.11	2.06	2.02	1.98	1.96	1.91	1.86	1.81	1.79	1.76	1.73	1.70	1.67	1.63

continued

$\alpha=0.10$

n_2 \ n_1	1	2	3	4	5	6	7	8	9	10	12	15	20	24	30	40	60	120	∞
20	2.97	2.59	2.38	2.25	2.16	2.09	2.04	2.00	1.96	1.94	1.89	1.84	1.79	1.77	1.74	1.71	1.68	1.64	1.61
21	2.96	2.57	2.36	2.23	2.14	2.08	2.02	1.98	1.95	1.92	1.87	1.83	1.78	1.75	1.72	1.69	1.66	1.62	1.59
22	2.95	2.56	2.35	2.22	2.13	2.06	2.01	1.97	1.93	1.90	1.86	1.81	1.76	1.73	1.70	1.67	1.64	1.60	1.57
23	2.94	2.55	2.34	2.21	2.11	1.05	1.99	1.95	1.92	1.89	1.84	1.80	1.74	1.72	1.69	1.66	1.62	1.59	1.55
24	2.93	2.54	2.33	2.19	2.10	2.04	1.98	1.94	1.91	1.88	1.83	1.78	1.73	1.70	1.67	1.64	1.61	1.57	1.53
25	2.92	2.53	2.32	2.18	2.09	2.02	1.97	1.93	1.89	1.87	1.82	1.77	1.72	1.69	1.66	1.63	1.59	1.56	1.52
26	2.91	2.52	2.31	2.17	2.08	2.01	1.96	1.92	1.88	1.86	1.81	1.76	1.71	1.68	1.65	1.61	1.58	1.54	1.50
27	2.90	2.51	2.30	2.17	2.07	2.00	1.95	1.91	1.87	1.85	1.80	1.75	1.70	1.67	1.64	1.60	1.57	1.53	1.49
28	2.89	2.50	2.29	2.16	2.06	2.00	1.94	1.90	1.87	1.84	1.79	1.74	1.69	1.66	1.63	1.59	1.56	1.52	1.48
29	2.89	2.50	2.28	2.15	2.06	1.99	1.93	1.89	1.86	1.83	1.78	1.73	1.68	1.65	1.62	1.58	1.55	1.51	1.47
30	2.88	2.49	2.28	2.14	2.05	1.98	1.93	1.88	1.85	1.82	1.77	1.72	1.67	1.64	1.61	1.57	1.54	1.50	1.46
40	2.84	2.44	2.23	2.09	2.00	1.93	1.87	1.83	1.79	1.76	1.71	1.66	1.61	1.57	1.54	1.51	1.47	1.42	1.38
60	2.79	2.39	2.18	2.04	1.95	1.87	1.82	1.77	1.74	1.71	1.66	1.60	1.54	1.51	1.48	1.44	1.40	1.35	1.29
120	2.75	2.35	2.13	1.99	1.90	1.82	1.77	1.72	1.68	1.65	1.60	1.55	1.48	1.45	1.41	1.37	1.32	1.26	1.19
∞	2.71	2.30	2.08	1.94	1.85	1.77	1.72	1.67	1.63	1.60	1.55	1.49	1.42	1.38	1.34	1.30	1.24	1.17	1.00

$\alpha=0.05$

n_2 \ n_1	1	2	3	4	5	6	7	8	9	10	12	15	20	24	30	40	60	120	∞
1	161.4	199.5	215.7	224.6	230.2	234.0	236.8	238.9	240.5	241.9	243.9	245.9	248.0	249.1	250.1	251.1	252.2	253.3	254.3
2	18.51	19.00	19.16	19.25	19.30	19.33	19.35	19.37	19.38	19.40	19.41	19.43	19.45	19.45	19.46	19.47	19.48	19.49	19.50
3	10.13	9.55	9.28	9.12	9.01	8.94	8.89	8.85	8.81	8.79	8.74	8.70	8.66	8.64	8.62	8.59	8.57	8.55	8.53
4	7.71	6.94	6.59	6.39	6.26	6.16	6.09	6.04	6.00	5.96	5.91	5.86	5.80	5.77	5.75	5.72	5.69	5.66	5.63
5	6.61	5.79	5.41	5.19	5.05	4.95	4.88	4.82	4.77	4.74	4.68	4.62	4.56	4.53	4.50	4.46	4.43	4.40	4.36
6	5.99	5.14	4.76	4.53	4.39	4.28	4.21	4.15	4.10	4.06	4.00	3.94	3.87	3.84	3.81	3.77	3.74	3.70	3.67
7	5.59	4.74	4.35	4.12	3.97	3.87	3.79	3.73	3.68	3.64	3.57	3.51	3.44	3.41	3.38	3.34	3.30	3.27	3.23
8	5.32	4.46	4.07	3.84	3.69	3.58	3.50	3.44	3.39	3.35	3.28	3.22	3.15	3.12	3.08	3.04	3.01	2.97	2.93
9	5.12	4.26	3.86	3.63	3.48	3.37	3.29	3.23	3.18	3.14	3.07	3.01	2.94	2.90	2.86	2.83	2.79	2.75	2.71

$\alpha=0.05$ continued

n_2 \ n_1	1	2	3	4	5	6	7	8	9	10	12	15	20	24	30	40	60	120	∞
10	4.96	4.10	3.71	3.48	3.33	3.22	3.14	3.07	3.02	2.98	2.91	2.85	2.77	2.74	2.70	2.66	2.62	2.58	2.54
11	4.84	3.98	3.59	3.36	3.20	3.09	3.01	2.95	2.90	2.85	2.79	2.72	2.65	2.61	2.57	2.53	2.49	2.45	2.40
12	4.75	3.89	3.49	3.26	3.11	3.00	2.91	2.85	2.80	2.75	2.69	2.62	2.54	2.51	2.47	2.43	2.38	2.34	2.30
13	4.67	3.81	3.41	3.18	3.03	2.92	2.83	2.77	2.71	2.67	2.60	2.53	2.46	2.42	2.38	2.34	2.30	2.25	2.21
14	4.60	3.74	3.34	3.11	2.96	2.85	2.76	2.70	2.65	2.60	2.53	2.46	2.39	2.35	2.31	2.27	2.22	2.18	2.13
15	4.54	3.68	3.29	3.06	2.90	2.79	2.71	2.64	2.59	2.54	2.48	2.40	2.33	2.29	2.25	2.20	2.16	2.11	2.07
16	4.49	3.63	3.24	3.01	2.85	2.74	2.66	2.59	2.54	2.49	2.42	2.35	2.28	2.24	2.19	2.15	2.11	2.06	2.01
17	4.45	3.59	3.20	2.96	2.81	2.70	2.61	2.55	2.49	2.45	2.38	2.31	2.23	2.19	2.15	2.10	2.06	2.01	1.96
18	4.41	3.55	3.16	2.93	2.77	2.66	2.58	2.51	2.46	2.41	2.34	2.27	2.19	2.15	2.11	2.06	2.02	1.97	1.92
19	4.38	3.52	3.13	2.90	2.74	2.63	2.54	2.48	2.42	2.38	2.31	2.23	2.16	2.11	2.07	2.03	1.98	1.93	1.88
20	4.35	3.49	3.10	2.87	2.71	2.60	2.51	2.45	2.39	2.35	2.28	2.20	2.12	2.08	2.04	1.99	1.95	1.90	1.84
21	4.32	3.47	3.07	2.84	2.68	2.57	2.49	2.42	2.37	2.32	2.25	2.18	2.10	2.05	2.01	1.96	1.92	1.87	1.81
22	4.30	3.44	3.05	2.82	2.66	2.55	2.46	2.40	2.34	2.30	2.23	2.15	2.07	2.03	1.98	1.94	1.89	1.84	1.78
23	4.28	3.42	3.03	2.80	2.64	2.53	2.44	2.37	2.32	2.27	2.20	2.13	2.05	2.01	1.96	1.91	1.86	1.81	1.76
24	4.26	3.40	3.01	2.78	2.62	2.51	2.42	2.36	2.30	2.25	2.18	2.11	2.03	1.98	1.94	1.89	1.84	1.79	1.73
25	4.24	3.39	2.99	2.76	2.60	2.49	2.40	2.34	2.28	2.24	2.16	2.09	2.01	1.96	1.92	1.87	1.82	1.77	1.71
26	4.23	3.37	2.98	2.74	2.59	2.47	2.39	2.32	2.27	2.22	2.15	2.07	1.99	1.95	1.90	1.85	1.80	1.75	1.69
27	4.21	3.35	2.96	2.73	2.57	2.46	2.37	2.31	2.25	2.20	2.13	2.06	1.97	1.93	1.88	1.84	1.79	1.73	1.67
28	4.20	3.34	2.95	2.71	2.56	2.45	2.36	2.29	2.24	2.19	2.12	2.04	1.96	1.91	1.87	1.82	1.77	1.71	1.65
29	4.18	3.33	2.93	2.70	2.55	2.43	2.35	2.28	2.22	2.18	2.10	2.03	1.94	1.90	1.85	1.81	1.75	1.70	1.64
30	4.17	3.32	2.92	2.69	2.53	2.42	2.33	2.27	2.21	2.16	2.09	2.01	1.93	1.89	1.84	1.79	1.74	1.68	1.62
40	4.08	3.23	2.84	2.61	2.45	2.34	2.25	2.18	2.12	2.08	2.00	1.92	1.84	1.79	1.74	1.69	1.64	1.58	1.51
60	4.00	3.15	2.76	2.53	2.37	2.25	2.17	2.10	2.04	1.99	1.92	1.84	1.75	1.70	1.65	1.59	1.53	1.47	1.39
120	3.92	3.07	2.68	2.45	2.29	2.17	2.09	2.02	1.96	1.91	1.83	1.75	1.66	1.61	1.55	1.50	1.43	1.35	1.25
∞	3.84	3.00	2.60	2.37	2.21	2.10	2.01	1.94	1.88	1.83	1.75	1.67	1.57	1.52	1.46	1.39	1.32	1.22	1.00

$\alpha = 0.025$

continued

n_2 \ n_1	1	2	3	4	5	6	7	8	9	10	12	15	20	24	30	40	60	120	∞
1	647.8	799.5	864.2	899.6	921.8	937.1	948.2	956.7	963.3	968.6	976.7	984.9	993.1	997.2	1001	1006	1010	1014	1018
2	38.51	39.00	39.17	39.25	39.30	39.33	39.36	39.37	39.39	39.40	39.41	39.43	39.45	39.46	39.46	39.47	39.48	39.40	39.50
3	17.44	16.04	15.44	15.10	14.88	14.73	14.62	14.54	14.47	14.42	14.34	14.25	14.17	14.12	14.08	14.04	13.99	13.95	13.90
4	12.22	10.65	9.98	9.60	9.36	9.20	9.07	8.98	8.90	8.84	8.75	8.66	8.56	8.51	8.46	8.41	8.36	8.31	8.26
5	10.01	8.43	7.76	7.39	7.15	6.98	6.85	6.76	6.68	6.62	6.52	6.43	6.33	6.28	6.23	6.18	6.12	6.07	6.02
6	8.81	7.26	6.60	6.23	5.99	5.82	5.70	5.60	5.52	5.46	5.37	5.27	5.17	5.12	5.07	5.01	4.96	4.90	4.85
7	8.07	6.54	5.89	5.52	5.29	5.12	4.99	4.90	4.82	4.76	4.67	4.57	4.47	4.42	4.36	4.31	4.25	4.20	4.14
8	7.57	6.06	5.42	5.05	4.82	4.65	4.53	4.43	4.36	4.30	4.20	4.10	4.00	3.95	3.89	3.84	3.78	3.73	3.67
9	7.21	5.71	5.08	4.72	4.48	4.23	4.20	4.10	4.03	3.96	3.87	3.77	3.67	3.61	3.56	3.51	3.45	3.39	3.33
10	6.94	5.46	4.83	4.47	4.24	4.07	3.95	3.85	3.78	3.72	3.62	3.52	3.42	3.37	3.31	3.26	3.20	3.14	3.08
11	6.72	5.26	4.63	4.28	4.04	3.88	3.76	3.66	3.59	3.53	3.43	3.33	3.23	3.17	3.12	3.06	3.00	2.94	2.88
12	6.55	5.10	4.47	4.12	3.89	3.73	3.61	3.51	3.44	3.37	3.28	3.18	3.07	3.02	2.96	2.91	2.85	2.79	2.72
13	6.41	4.97	4.35	4.00	3.77	3.60	3.48	3.39	3.31	3.25	3.15	3.05	2.95	2.89	2.84	2.78	2.72	2.66	2.60
14	6.30	4.86	4.24	3.89	3.66	3.50	3.38	3.29	3.21	3.15	3.05	2.95	2.84	2.79	2.73	2.67	2.61	2.55	2.49
15	6.20	4.77	4.15	3.80	3.58	3.41	3.29	3.20	3.12	3.06	2.96	2.86	2.76	2.70	2.64	2.59	2.52	2.46	2.40
16	6.12	4.69	4.08	3.73	3.50	3.34	3.22	3.12	3.05	2.99	2.89	2.79	2.68	2.63	2.57	2.51	2.45	2.38	2.32
17	6.04	4.62	4.01	3.66	3.44	3.28	3.26	3.06	2.98	2.92	2.82	2.72	2.62	2.56	2.50	2.44	2.38	2.32	2.25
18	5.98	4.56	3.95	3.61	3.38	3.22	3.10	3.01	2.93	2.87	2.77	2.67	2.56	2.50	2.44	2.38	2.32	2.26	2.19
19	5.92	4.51	3.90	3.56	3.33	3.17	3.05	2.96	2.88	2.82	2.72	2.62	2.51	2.45	2.39	2.33	2.27	2.20	2.13
20	5.87	4.46	3.86	3.51	3.29	3.13	3.01	2.91	2.84	2.77	2.68	2.57	2.46	2.41	2.35	2.29	2.22	2.16	2.09
21	5.83	4.42	3.82	3.48	3.25	3.09	2.97	2.87	2.80	2.73	2.64	2.53	2.42	2.37	2.31	2.25	2.18	2.11	2.04
22	5.79	4.38	3.78	3.44	3.22	3.05	2.73	2.84	2.76	2.70	2.60	2.50	2.39	2.33	2.27	2.21	2.14	2.08	2.00
23	5.75	4.35	3.75	3.41	3.18	3.02	2.90	2.81	2.73	2.67	2.57	2.47	2.36	2.30	2.24	2.18	2.11	2.04	1.97
24	5.72	4.32	3.72	3.38	3.15	2.99	2.87	2.78	2.70	2.64	2.54	2.44	2.33	2.27	2.21	2.15	2.08	2.01	1.94

continued

$\alpha=0.025$

n_2 \ n_1	1	2	3	4	5	6	7	8	9	10	12	15	20	24	30	40	60	120	∞
25	5.69	4.29	3.69	3.35	3.13	2.97	2.85	2.75	2.68	2.61	2.51	2.41	2.30	2.24	2.18	2.12	2.05	1.98	1.91
26	5.66	4.27	3.67	3.33	3.10	2.94	2.82	2.73	2.65	2.59	2.49	2.39	2.28	2.22	2.16	2.09	2.03	1.95	1.88
27	5.63	4.24	3.65	3.31	3.08	2.92	2.80	2.71	2.63	2.57	2.47	2.36	2.25	2.19	2.13	2.07	2.00	1.93	1.85
28	5.61	4.22	3.63	3.29	3.06	2.90	2.78	2.69	2.61	2.55	2.45	2.34	2.23	2.17	2.11	2.05	1.98	1.91	1.83
29	5.59	4.20	3.61	3.27	3.04	2.88	2.76	2.67	2.59	2.53	2.43	2.32	2.21	2.15	2.09	2.03	1.96	1.89	1.81
30	5.57	4.18	3.59	3.25	3.03	2.87	2.75	2.65	2.57	2.51	2.41	2.31	2.20	2.14	2.07	2.01	1.94	1.87	1.79
40	5.42	4.05	3.46	3.13	3.90	2.74	2.62	2.53	2.45	2.39	2.29	2.18	2.07	2.01	1.94	1.88	1.80	1.72	1.64
60	5.29	3.93	3.34	3.01	2.79	2.63	2.51	2.41	2.33	2.27	3.17	2.06	1.94	1.88	1.82	1.74	1.67	1.58	1.48
120	5.15	3.80	3.23	2.89	2.67	2.52	2.39	2.30	2.22	2.16	2.05	1.94	1.82	1.76	1.69	1.61	1.53	1.43	1.31
∞	5.02	3.69	3.12	2.79	2.57	2.41	2.29	2.19	2.11	2.05	1.94	1.83	1.71	1.64	1.57	1.48	1.39	1.27	1.00

$\alpha=0.01$

n_2 \ n_1	1	2	3	4	5	6	7	8	9	10	12	15	20	24	30	40	60	120	∞
1	4052	4999.5	5403	5625	5764	5859	5928	5982	6022	6056	6106	6157	6209	6235	6261	6287	6313	6339	6366
2	98.50	99.00	99.17	99.25	99.30	99.33	99.36	99.37	99.39	99.40	99.42	99.43	99.45	99.46	99.47	99.47	99.48	99.49	99.50
3	34.12	30.82	29.46	28.71	28.24	27.91	27.67	27.49	27.35	27.23	27.05	26.87	26.69	26.60	26.50	26.41	26.32	26.22	26.13
4	21.20	18.00	16.69	15.98	15.52	15.21	14.98	14.80	14.66	14.55	14.37	24.20	14.02	13.93	13.84	13.75	13.65	13.56	13.46
5	16.26	13.27	12.06	11.39	10.97	10.67	10.46	10.29	10.16	10.05	9.89	9.72	9.55	9.47	9.38	9.29	9.20	9.11	9.02
6	13.75	10.93	9.78	9.15	8.75	8.47	8.26	8.10	7.98	7.87	7.72	7.56	7.40	7.31	7.23	7.14	7.06	6.97	6.88
7	12.25	9.55	8.45	7.85	7.46	7.19	6.99	6.84	6.72	6.62	6.47	6.31	6.16	6.07	5.99	5.91	5.82	5.74	5.65
8	11.26	8.65	7.59	7.01	6.63	6.37	6.18	6.03	5.91	5.81	5.67	5.52	5.36	5.28	5.20	5.12	5.03	4.95	4.86
9	10.56	8.02	6.99	6.42	6.06	5.80	5.61	5.47	5.35	5.26	5.11	4.96	4.81	4.73	4.65	4.57	4.48	4.40	4.31

$\alpha = 0.01$

continued

n_2 \ n_1	1	2	3	4	5	6	7	8	9	10	12	15	20	24	30	40	60	120	∞
10	10. 04	7. 56	6. 55	5. 99	5. 64	5. 39	5. 20	5. 06	4. 94	4. 85	4. 71	4. 56	4. 41	4. 33	4. 25	4. 17	4. 08	4. 00	3. 91
11	9. 65	7. 21	6. 22	5. 67	5. 32	5. 07	4. 89	4. 74	4. 63	4. 54	4. 40	4. 25	4. 10	4. 02	3. 94	3. 86	3. 78	3. 69	3. 60
12	9. 33	6. 93	5. 95	5. 41	5. 06	4. 82	4. 64	4. 50	4. 39	4. 30	4. 16	4. 01	3. 86	3. 78	3. 70	3. 62	3. 54	3. 45	3. 36
13	9. 07	6. 70	5. 74	5. 21	4. 86	4. 62	4. 44	4. 30	4. 19	4. 10	3. 96	3. 82	3. 66	3. 59	3. 51	3. 43	3. 34	3. 25	3. 17
14	8. 86	6. 51	5. 56	5. 04	4. 69	4. 46	4. 28	4. 14	4. 03	3. 94	3. 80	3. 66	3. 51	3. 43	3. 35	3. 27	3. 18	3. 09	3. 00
15	8. 68	6. 36	5. 42	4. 89	4. 56	4. 32	4. 14	4. 00	3. 89	3. 80	3. 67	3. 52	3. 37	3. 29	3. 21	3. 13	3. 05	2. 96	2. 87
16	8. 53	6. 23	5. 29	4. 77	4. 44	4. 20	4. 03	3. 89	3. 78	3. 69	3. 55	3. 41	3. 26	3. 18	3. 10	3. 02	2. 93	2. 84	2. 75
17	8. 40	6. 11	5. 18	4. 67	4. 34	4. 10	3. 93	3. 79	3. 68	3. 59	3. 46	3. 31	3. 16	3. 08	3. 00	2. 92	2. 83	2. 75	2. 65
18	8. 29	6. 01	5. 09	4. 58	4. 25	4. 01	3. 94	3. 71	3. 60	3. 51	3. 37	3. 23	3. 08	3. 00	2. 92	2. 84	2. 75	2. 66	2. 57
19	8. 18	5. 93	5. 01	4. 50	4. 17	3. 94	3. 77	3. 63	3. 52	3. 43	3. 30	3. 15	3. 00	2. 92	2. 84	2. 76	2. 67	2. 58	2. 49
20	8. 10	5. 85	4. 94	4. 43	4. 10	3. 87	3. 70	3. 56	3. 46	3. 37	3. 23	3. 09	2. 94	2. 86	2. 78	2. 69	2. 61	2. 52	2. 42
21	8. 02	5. 78	4. 87	4. 37	4. 04	3. 81	3. 64	3. 51	3. 40	3. 31	3. 17	3. 03	2. 88	2. 80	2. 72	2. 64	2. 55	2. 46	2. 36
22	7. 95	5. 72	4. 82	4. 31	3. 99	3. 76	3. 59	3. 45	3. 35	3. 26	3. 12	2. 98	2. 83	2. 75	2. 67	2. 58	2. 50	2. 40	2. 31
23	7. 88	5. 66	4. 76	4. 26	3. 94	3. 71	3. 54	3. 41	3. 30	3. 21	3. 07	2. 93	2. 78	2. 70	2. 62	2. 54	2. 45	2. 35	2. 26
24	7. 82	5. 61	4. 72	4. 22	3. 90	3. 67	3. 50	3. 36	3. 26	3. 17	3. 03	2. 89	2. 74	2. 66	2. 58	2. 49	2. 40	2. 31	2. 21
25	7. 77	5. 57	4. 68	4. 18	3. 85	3. 63	3. 46	3. 32	3. 22	3. 13	2. 99	2. 85	2. 70	2. 62	2. 54	2. 45	2. 36	2. 27	2. 17
26	7. 72	5. 53	4. 64	4. 14	3. 82	3. 59	3. 42	3. 29	3. 18	3. 09	2. 96	2. 81	2. 66	2. 58	2. 50	2. 42	2. 33	2. 23	2. 13
27	7. 68	5. 49	4. 60	4. 11	3. 78	3. 56	3. 39	3. 26	3. 15	3. 06	2. 93	2. 78	2. 63	2. 55	2. 47	2. 38	2. 29	2. 20	2. 10
28	7. 64	5. 45	4. 57	4. 07	3. 75	3. 53	3. 36	3. 23	3. 12	3. 03	2. 90	2. 75	2. 60	2. 52	2. 44	2. 35	2. 26	2. 17	2. 06
29	7. 60	5. 42	4. 54	4. 04	3. 73	3. 50	3. 33	3. 20	3. 09	3. 00	2. 87	2. 73	2. 57	2. 49	2. 41	2. 33	2. 23	2. 14	2. 03
30	7. 56	5. 39	4. 51	4. 02	3. 70	3. 47	3. 30	3. 17	3. 07	2. 98	2. 84	2. 70	2. 55	2. 47	2. 39	2. 30	2. 21	2. 11	2. 01
40	7. 31	5. 18	4. 31	3. 83	3. 51	3. 29	3. 12	2. 99	2. 89	2. 80	2. 66	2. 52	2. 37	2. 29	2. 20	2. 11	2. 02	1. 92	1. 80
60	7. 08	4. 98	4. 13	3. 65	3. 34	3. 12	2. 95	2. 82	2. 72	2. 63	2. 50	2. 35	2. 20	2. 12	2. 03	1. 94	1. 84	1. 73	1. 60
120	6. 85	4. 79	3. 95	3. 48	3. 17	2. 96	2. 79	2. 66	2. 56	2. 47	2. 34	2. 19	2. 03	1. 95	1. 86	1. 76	1. 66	1. 53	1. 38
∞	6. 63	4. 61	3. 78	3. 32	3. 02	2. 80	2. 64	2. 51	2. 41	2. 32	2. 18	2. 04	1. 88	1. 79	1. 70	1. 59	1. 47	1. 32	1. 00

References

[1] Walpole R E，Myers R H，Myers S L，et al. Probability & Statistics for Engineer & Scientists. 9th ed. New York：Prentice Hall，2011.

[2] Devore J L. Probability and Statistics for Engineering and the Sciences. Ninth Edition. Boston：Cengage Learning，2014.

[3] 赖虹建，郝志峰. Probability and Statistics. 北京：高等教育出版社，2008.

[4] 盛骤，谢式千，潘承毅. 概率论与数理统计. 4 版. 北京：高等教育出版社，2008.